WILD
IDEAS

WILD
IDEAS

David Rothenberg
editor

 University of Minnesota Press
Minneapolis
London

Published by the University of Minnesota Press
111 Third Avenue South, Suite 290
Minneapolis, MN 55401-2520
Printed in the United States of America on acid-free paper

Library of Congress Cataloging-in-Publication Data

Wild ideas / edited by David Rothenberg.
    p. cm.
    Includes the results of the 5th World Wilderness Congress held in
Tromsø, Norway in September 1993.
    Includes index.
    ISBN 0-8166-2614-6 (hc). — ISBN 0-8166-2615-4 (pb)
    1. Nature conservation—Philosophy—Congresses. 2. Ecology—Philosophy—
Congresses. 3. Wilderness areas—Congresses. 4. Wildlife conservation—Philosophy—
Congresses. I. Rothenberg, David, 1962–   . II. World Wilderness Congress
(5th : 1993 : Tromsø, Norway)
QH75.A1W528   1995
333.7'2'01—dc20   94-49620

The University of Minnesota is an equal-opportunity educator and employer.

Contents

Illustrations

Foreword

In the arena of international nature conservation, no concept is fraught with more differences in personality, policy, and particularity than is *wilderness*. Yet, is there anything more important? Wilderness is our original home, our genetic mother, our biological lifeline, our earthly father, and our future security. It is both the bosom that nourishes us and the stick that ultimately corrects our misbehavior.

Therefore, the World Wilderness Congress (WWC) was founded in 1977 upon the understanding that a healthy and sane human society is impossible without wilderness, and that the answer to saving wilderness—for our security or for its own sake—lies in nations, people, and professions cooperating together. The WWC's rationale is, simply, that wilderness conservation must integrate cultural, professional, and political differences. To integrate such differences, one must first listen to his or her colleagues, friends, and adversaries, understanding where they are coming from, what they are made of, and in what direction they are heading. From this integration emerges effective action.

Hence the critical importance of wilderness philosophy to international nature conservation. It helps clarify our origins, establish our identity, discern our various relations to nature, and set our direction. Without this philosophical core, international conservation policy will be ineffective, and shift with the vagaries of politics while held captive by the trepidation of economics.

As a result, the WWC has always placed philosophy on an equal footing with science, management, economics, and education, and has helped establish credibility for a holistic approach to environmental problem solving. Through five meetings over sixteen years—in South Africa (1977), Australia (1980), Scotland (1983), the United States (1987), and Norway (1993)—the WWC has become a unique forum and platform for international action in nature conservation, allowing the public to meet and work with professionals from all sectors, from game rangers to politicians, scientists to poets, managers to philosophers.

At the fifth WWC in Tromsø, Norway, in September 1993, David Rothenberg convened an unusual and talented group of thinkers in a symposium to investigate the idea of the wild. In the resulting collection, *Wild Ideas*, he

brings to light the best of their presentations and discussions, along with related contributions from other authors, thereby illuminating the philosophical perspectives that lie at the center of humankind's relationship to wild
nature, clarifying the critical thinking that drives us to protect or destroy
wilderness, ultimately asking the question: Do we understand wildness, or
know enough to realize it is something large enough to always remain a few
steps beyond understanding?

Vance G. Martin, director
World Wilderness Congress
The WILD Foundation
Ojai, California

Introduction
Wildness Untamed:
The Evolution of an Ideal

David Rothenberg

What would the world be, once bereft
Of wet and of wilderness? Let them be left,
O let them be left, wildness and wet:
Long live the weeds and the wilderness yet.
Gerard Manley Hopkins

There are at least two ways to sing the praises of wilderness. One is Hopkins's rhythmic paean to the dank and the squishy muck of the earth in its useless beauty dares us to leave it alone so that we will not sanitize our home out of existence. On the other hand, there is Chip Taylor's much-covered popular song, "Wild Thing," which tries to bring down to earth the allure, the guts of romanticism, the satisfaction of wildness making our hearts sing right in the human midst. Wildness as that other least human realm of nature, to love and respect in its independence or necessary distance, and also the craziness we fear and yet need right in our world, among people, in our crazy culture, in the excesses of poetry or the whole industrial mess of our sublime time (sublime in the extreme edginess of its beauty, in our abhorrent fascination with what goes on in and among us, and then in that bounded place apart).

In this collection I have tried to gather essays that challenge our notions of wilderness which have been so central to the development of conservation and environmentalism in this country and also the world. The idea of wildness has always been deeper than a place that exists because we place boundaries around it and call it "a wilderness." We love and fear the wild in all parts of our culture, and it exerts its power because of this. Here we investigate the extent of this elusive and alluring quality, and examine its implications for the notions of human and nonhuman nature and the future of environmentalism in thought and practice.

As ecological problems receive more serious scrutiny from both the government and the public, there seems to be a growing discontent or confusion surrounding the idea of wilderness. Wilderness as idea has been essential to the furtherance of civilization. First it was dangerous expanse beyond the frontier that needed to be tamed in order to be ready for human habitation. Now, as the world has been mowed over by our renovations, nostalgia for the natural has set in. We long for the wilderness. We pine for the shrinking grandeur of nature untrammeled by humankind. Its value to us seems to escalate beyond belief. "Bully!" intones Teddy Roosevelt at the brink of the Grand Canyon. "Leave it as it is." Still, we push our culture as close to the edge of the sublime expanse as we can. Sixty years later, Floyd Dominy talks about damming the innards out of the place so that motorboaters can power swiftly up the inner gorge and admire the greatness from within, not a drop of sweat spent on the struggle to get there.[1] Why worry? No one would *see* the damage from above. Yet we would know it is there. A crime against the wild would have been committed.

That's old news now, nearly a quarter century into the haze of memory. We have left the inside of the Canyon more or less alone. Still, the great tragedy of places like Glen Canyon, north up into southeastern Utah, was one of the saddest and most moving tales of my childhood. I grew up on those Sierra Club picture books such as *The Place No One Knew*, and the beauty of Music Temple was all the more graphic because I read that these places were now under hundreds of feet of water. The lost wilderness was even more touching than what still could be saved. I ran outside, climbed a tree, resolved to devote my life to the saving of the wild.

A fine dream—but how much does the pristine have to do with the environmental problems that plague our world today? The telling toxicity of our culture becomes all the more apparent as we produce wastes that have nowhere to go, piling up unclaimed and unmanaged throughout the world, where only the ignorant or the poor can agree to accept them. The atmosphere heats up, and the changes are unforeseeable. The wilderness we once took for granted is now being swallowed up by the "progress" that taught us to recognize the wild as a good in the first place.

The wild is not to be feared anymore, but to be loved, savored, enjoyed, inhabited respectfully so that we may ensure its presence long into the future, for seven or seventeen generations hence into the unknown. The enemy is us now, and if we change ourselves, the world may endure with us in it.

These seem like old platitudes to me today, but they can be repeated again and again. In this multicultural world that intellectual life now encourages, it is easier to disagree and to shoot down generalized concepts than it is to agree to take action. Who is wilderness for? Just those rich enough to have leisure time to enjoy it? All of us who decide it's worth keeping even if it might cloak millions in mineral or oil resources? Is it for everyone, even those who choose not to recognize its worth? Or is it for the Earth itself, an intrinsic value that humans should respect toward a future in which we see the world's interests before our own?

Certainly wilderness has guided the conservation movement in this country and helped it evolve into the environmental movement that is gaining wider acceptance all the time. But is wilderness an ethnocentric concept that has little to do with the more profound and direct ways in which nature is experienced by the world's peoples—nature as the fabric of material culture, as the source of sustenance practical and aesthetic, lived with as home, not as any place empty of human community? The idea of wilderness, I believe, is an invention of civilization, a foil for our failures and our successes. It is important, certainly, but maybe more regionally than globally. Or as part of an environmental philosophy, not a whole.

The idea of wilderness may alienate those who live close with nature, the pastoralists, the farmers, the contented suburbanites. It is still the beautiful shock to those of us in the most extreme of cities, who find silence to be the exception, not the rule.

Wilderness is not everyone's idea of nature. Nor should it be. Nor should it be the "best" kind of nature to which all other plans for recovery aspire. It is one kind of sacred space we in our built-up environment have chosen to christen with value. It is worth more than we can say. To us. Now. We want to legislate that value into something eternal, or at least into something that will last.

Among philosophers there has recently been quite some debate as to the relevance of wilderness to environmental understanding. Has it been the mainstream guiding idea behind increasing human respect for nature, or is it one rather extremist notion? The leading ecophilosophers Baird Callicott and Holmes Rolston have taken their debate on the road, and one runs across them again and again sitting on opposite sides of the fence. Callicott has his hero Aldo Leopold agreeing with Aristotle in believing that humanity completes what nature has begun. In his now legendary article "The Wilderness Idea Revisited,"[2] Callicott argues that wilderness preservation is limited, shortsighted, Americocentric, and encouraging to a philosophy that separates even an enlightened humanity from the nature we wish to save. He argues instead for a sensitive kind of sustainable development, where human culture is meant to progress and flourish in tandem with nature, improving biodiversity, managing the Earth into better balance.

It is, in some senses, a surprising position for the major interpreter of Aldo Leopold to take. Leopold is famous for his land ethic, which stated that "a thing is right when it tends to preserve the integrity, stability, and beauty of the biotic community. It is wrong when it tends otherwise."[3] It does not say that a thing is right if it *improves* the biotic community. Nevertheless, wilderness advocates lately have been accused of wanting to stop time, to hold back change, while at the same time the science of ecology has gone on to embrace a more dynamic notion of what nature is and will be. Callicott is only providing a mediating response for our time, hoping a reasonable compromise might be reached.

Holmes Rolston, in his response, notes that it is just in such questioning times like these that we need a firm wilderness idea in tow.[4] We may be a part of nature, but what we are part of remains larger than us, and deserves deference and enough respect so that we may let it flourish unimpeded—if we can, given that so much damage has already been done. Who could be against the preservation of wilderness, now that so little is left? he asks. Love of wilderness does have a religious, and hence nonrational, aspect, but it need not be dichotomized and institutionalized as such. (It is always easier to make opposing positions seem more extreme than they are. This is the perennial appeal of the "straw man" argument—make your opponent into something you can beat with logic. You come out the winner, but you may shy from the wild ambiguity of truth.) No one is saying that love of wilderness implies hate for civilization. Indeed, civilization is, in my opinion, the precursor for the identification and then quest for wilderness that seems less attainable with every year. Wilderness is necessary, says Rolston, and needs more solid protection *at the same time as the idea of wilderness receives increased criticism*:

> We do not want the whole Earth without civilization, for we believe that humans belong on Earth; Earth is not whole without humans and their civilization, without the political animal building his *polis*, without peoples inheriting their promised lands. Civilization is a broken affair, and in the long struggle to make and keep life human, moral, even godly, perhaps there should be islands, sanctuaries of moral goodness within a civilization often sordid enough. But that is a different issue from whether, when we build our civilizations for better or worse, we also want to protect where and as we can those nonhuman values in wild spontaneous nature that preceded and yet surround us.[5]

Is it so different? This whole dispute seems at one level neither here nor there. Of course we should have more wilderness; of course wilderness is not all there is to the saving of nature.

But the wild and the spontaneous are more than values independent of human aspiration. They may apply to *us* as well. If we are to learn what inhabitation of the wild and the immediate might be, we need to expand our understanding of the notion of wild as attribute, not place, into different ways of writing and speaking, different cultures, different problems. This expansion is the task of this book, in which I have brought together writings that try to set the very notion of the wild free into untamed places where wildness as quality may hope to enliven and explain things that have been bowled over by too many rules or by straw-dog attempts of one side's logic to outbark the other's.

The *wild* is more than a named place, an area to demarcate. It is a quality that beguiles us, a tendency we both flee and seek. It is the unruly, that which won't be kept down, that crazy love, that path that no one advises us to take—it's against the rules, it's too far, too fast, beyond order, irreconcilable with what we are told is right. Wild Thing. Wild Life. Wild One. Wild Child. Wild Culture. You make my heart sing. But who knows what tomorrow will bring?

The wild refers to many things. In wildness is the preservation of the world, says our canonized curmudgeon Thoreau. He did not say "wilderness." He did not mean wilderness. He meant the breaking of rules, the ostracized life in the midst of his peers. Walden Pond is a mile from downtown Concord, and a train runs close to the far shore today just as it did back in Henry David's time. That nearness is the wild in it. To buck civilization right in its midst. He said the wild is all that interests us in literature. That's where he lures us away from losing ourselves in nature to finding a deep surge of nature far inside us, what he named as the soul force behind creativity. Poet Philip Booth echoes the sentiment: "whether we live and write in sight of Mt. Rainier or in midtown Manhattan, no matter where we experience being in place, we immerse in our deepest selves when we begin to write. It's from instinctive memory, from *the wilderness of the imagination,* from *a mindfulness forever wild,* that Art starts."[6]

Once again, Thoreau found the wild close to the tame, and pledged allegiance to it as much in his bucking of authority as in any spurning of culture's ways. It is probably on the summit of Maine's Mount Katahdin that his writing, usually ornate, mannered, nearly Victorian, approaches a free wildness that approaches the timeless, animistic sense of wild wanderers from many cultures. He is here describing the windswept, barren summit upon which almost no visible plant grows, rising high above dense forests:

> It [is] a place . . . to be inhabited by men nearer of kin to the rocks and to wild animals than we . . . I stand in awe of my body, this matter to which I am bound has become so strange to me. I fear not spirits, ghosts, of which I am one—*that* my body might—but I fear bodies, I tremble to meet them. What is this Titan that has taken possession of me? Talk of mysteries—Think of our life in nature—daily to be shown matter, to come in contact with it—rocks, trees, wind on our cheeks! the *solid* earth! the *actual* world! the *common sense*! Contact! Contact!

He revels in the aloneness of the surrounding place, an enveloping world of raw natural phenomena. Reaching through appearance, Thoreau has touched the Earth and is groping for the language to say this. His cranky style fails him, his ruminations on wildness as tonic have no place. He is in the element here, and it nearly makes him speechless. Go there yourself; you will feel what he means. The wild barrenness of such a mountain is chillingly inhuman, but at the same time touches us deep inside our own human selves, as a weight in the gut somehow proves that the wild within us belongs here, and will not forget the place whatever home we choose.

So take people out there. Get them to experience the wild firsthand, and then they will know, love, and demand that it be preserved for ourselves, future generations, and earthly integrity itself. That was the message of John Muir, already almost a hundred years ago: "Thousands of tired, nerve-shaken, over-civilized people are beginning to find out that going to the mountains is going home; that wildness is a necessity; and that mountain parks and reservations are useful not only as fountains of timber . . . , but as fountains of life."[7] So he begat the Sierra Club, which swelled exponentially in membership like every other indicator of human progress during the twentieth century. Philosopher-mountain guide Jack Turner has called this statement "Muir's Mistake" because Muir did not realize that the experience of the wild could *never* be duplicated by a Sierra Club trip, or by any organizational structure that defines and demarcates the wild, taking you there, setting up boundaries to say what is wild and what is not. The naming of wilderness is part of its commodification.[8] We in the grip of the commercialization of nature had best be extra careful about what we imagine wilderness to mean, and what we use the concept to demonstrate. Throughout its popular history, the idea of wilderness has shown itself time and again to be the creation of human consciousness, malleable in the extreme, used to fulfill our deepest desires and worst fears.

Reading through Roderick Nash's epic and oft-revised *Wilderness and the American Mind,* I am struck by how little has changed in the way we talk about wilderness and imagine it to be so many things. Apocalyptic visions of nature ending under human wrath were around long before we possessed the technology to blow ourselves to smithereens. Wild adventurers have regaled armchair readers as long as there have been frontiers to fantasize about.

I am intrigued by the story of Joseph Knowles, the wild media sensation of his generation, who in 1913 "disrobed in a cold drizzle at the edge of a lake in northeastern Maine, smoked a final cigarette, shook hands around a group of sportsmen and reporters, and trudged off into the wilderness."[9] Carrying no equipment of any kind, he headed out into the bush to live pure and untrammeled, living off the land. Two months later he returned to champion the values of the primal life, and he became a national celebrity, parading through the Boston streets and selling 300,000 copies of his book, *Alone in the Wilderness,* in no time. Never mind that one muckraking newspaper tried to prove that Knowles was a fraud who had spent most of the time in a secure and dry hut; Americans desperately wanted to believe in the plausibility of a current and noble savage. Wilderness has since that time been something of a national cult, an American ideal. And it is no surprise that, along with all those other ideals, the idea of wilderness today is being exported all over the globe, not without serious resistance, some from multicultural environmentalists.

At the fifth World Wilderness Congress in arctic Norway, at which many of these essays originated, there was a general sense that the notion of wilderness was imperialistic, a sense of luxury developed in the northern countries, particularly the United States. What follows is my reading of the sentiments of the whole group involved in the "Idea of the Wild" symposium, both panelists and respondents from the audience.

We heard local northern representatives complain that wilderness is a foreign concept to them, set up by "urban people" who want to spend only their free time in nature. We heard native people demand that they not be pigeonholed or stereotyped so that they cannot enjoy the pleasures of development that the rest of us have experienced enough so we can see their problems. We have heard others speak of the need for wilderness to be experienced firsthand, or, as Peter Matthiessen wrote of the elusive snow leopard, it may be enough for some to know that wilderness exists and is guaranteed in this world, without the need to ever see it.[10]

One of the few points of agreement in our seminar was the sense that people mean many different things when they speak of wilderness, or of the wild. In Canada, wilderness usually includes native people, whereas in the

United States, they are often excluded. In Nepal, there is hardly any wilderness below five thousand meters, because people have lived close to the land for centuries. We heard this report on the idea of wilderness in South Africa:

> In Zulu, the word *uchhlati* is the word for wilderness, and it actually means tree. The impacts on wilderness are related to the numbers of people in that landscape, and the sense that people cannot really increase in population densities. Diseases were limiting populations. There was no need to name wilderness because the populations were one person in a thousand kilometers. But now you have a fairly rapid change in population densities, which will impact on their very survival. We're talking today about wilderness in temperate areas, but in the subtropics, we are faced with complete change. The old Zulu word may not be enough for the Zulus. The developing countries are getting homogenized, but wilderness still has a practical role to play, and it is easy to get lost in philosophical discussion. You're talking about pears and I'm talking about bananas.[11]

We need to recognize the differences in our understanding of the term before we try to reach some consensus in how to live successfully in wild places. We opposed the term *sustainable development,* which is so slippery as to be usable to mean sustaining just about any practice the promoter wants to sustain or develop. We affirmed the right of local people to be given the chance to manage their own local resources; however, they, like everyone else, must show that their specific knowledge builds on what the world as a whole now knows about itself. The old ways should not continue just because they are the old ways. We cannot let support for local cultures be put forth as a disguise for a veiled nationalism. Nationalism has no place at all in the arena of international environmental issues.

Enough negativity. So what then did the members of our session support? We are for a "wild culture," where nature is not opposed to humanity, where use of the surrounding world does not require calling it "resources" and us, "consumers." We spoke of classical and romantic notions of wilderness—the first naming the wilderness as the home of the fearful and evil, and the second, as the place where we come to our senses out of the madness of civilization. Neither is appropriate any longer. The wild will transcend the tame and the untamed, the jungle and the city. It must refer to our progress toward a culture that can understand nature without hemming it in by management and control. The wild should not be bad and unruly, but we should not rely on too many rules to get there. The word *wild* is a challenge for us to conceive a new kind of civilization, one that does not require the destruction of our world to improve itself. We will not find the model for this culture in the past or in the present. Only in the future will we get there.

This conference conjoined the issues of wilderness and development, assuming that they fit neatly together. Well, I have met too many people who are either for destroying the wilderness, or for saving it, with little chance of communication between them. Take away wilderness as an opposition to civilization, think instead about the wild, and see if we can communicate. Take people out there. Listen to the silence. Find a way to live in and with it. We in the developed world must *change our way of life*. Consume less. Not because someone tells us to, but because we want to. To sense the wild. To live more for each other and the nature that lies around us.

Then enough about people. The wild is not just for us. It is for the non-human, or the more than human. It is for the majority of the world that does not have to live in the wasteful and peculiar manner into which we have gotten ourselves. The wild is for the trees, the fish, the wolves, the bear, the mosquito, the cloudberry, the water, and the waves. Our ideas need to encompass the kind of care that is generous enough to care for more than humanity alone. Wildness is a deep intellectual and passionate quest, demanding new effort in our restructuring of thought, expression, and ways of writing. I have encouraged all of these here. Question authority, especially ideological authority, methodological authority, and the truisms of ecological correctness.

I want the guide of the wild to lead me to see, to touch with all senses the depth that is nature wherever I am. The city, too, must seem to breathe life as its concrete and steel thrust too far into the air to keep the horizon alive. I am desperate to see a tree, and that is not for nostalgia alone. I have retreated into the realm of words, and these in a way seem tame no matter what savage bewilderment is applied to them. Wasteland and thicket, arena of decay and destroyed emptiness, fearsome and strange. A growl, a howl, a scream. Wile E. Coyote: Super Genius. *Wild Kingdom*, brought to you by an insurance company in Nebraska. Assurance that all is okay, though we know not what will happen. The wild rages on, there is placidity and violence there. We want to get along but then turn against each other, again and again. Does the wild encourage such conflict or spurn it?

We see what we want out there in the woods or desert, in the nature film of our minds, taking rough cuts and turning them into manageable stories for the material memories of other species' worlds. And wildness insists there is always more out there than we will be able to see, or be able to explain.

Now, what does each essay contribute to this search for redefinition and understanding?

Whose Wild Idea?

The first three essays present an overview of present conceptions of the idea of wilderness, its problems, and its possible futures. Ed Grumbine begins the book with a superb review of how wilderness is currently understood by conservation professionals, suggesting ways in which the debate could continue to be refined. He calls wildness a deeper process than wilderness, something that matters more, that is harder to legislate, a flowing activity with a deep uneasiness in relation to our culture. I would say that it is also more daring, more ambiguous an idea, as much in human order as in natural law.

Grumbine reviews the Rolston/Callicott debate on the efficacy of wilderness as an international conservation imperative, and warns of the hidden political agenda in the Wise Use Movement (WUM). He then discusses what would be necessary to make wilderness part of the emerging doctrine of sustainable development, and begins to speculate on what will be needed to ready the idea of the wild for the future.

Next, geographer Denis Cosgrove writes from England as a critic of the sheer Americanness of the wilderness idea. He defines wilderness as the place where urbanity always ends, but he wonders just who are the people attracted to such a metaphorical place today. Is it just the original, "pure," white-bread Americans, who know the elaborate rules of conduct, the ways of being, who think it is fun? Is it for all of us, or only the privileged? What kind of immigrants to America are welcome in this recreationally defined wilderness?

Wilderness insiders may find this piece somewhat alarmist, but I believe it is a necessary tonic to an enthusiasm that sometimes becomes naive in the wake of the diverse realities of the rest of the world. Europe has no native doctrine of wilderness, and even the far reaches of Scandinavia are more comfortable with the notion of "free nature" open to all, but not named by the troubling term *wild*.[12] Cosgrove offers a sobering history of factual data relating to the development of the wilderness ideal, revealing it to be one of many possible cultural interpretations of what we desire in nature.

Max Oelschlaeger takes off from his seminal *Idea of Wilderness* to offer a rousing plea for the reform of language and culture to bring them closer to the Earth. He has become more outspoken and radical with time. Now against sustainable development, Max argues instead for a renovated sense of "wild culture," beyond environmentalism, open to a whole new way of speaking about our place in the world. He names it "Earth-talk," and it is up to us to figure out what this means. Oelschlaeger questions himself, wonders what words can accomplish, questions what literacy can do for the wild. (Later essays by Abram, Rothenberg, and Schafer offer some possible clues.)

Cross-Cultural Wild

The next three chapters are concerned with the translation of the wilderness idea across cultures. Marvin Henberg tackles the question of whether wilderness, as a philosophic term, can be exported across the globe. He is optimistic and believes that the wilderness ideal will become palatable among many cultures—that it must in order to evolve into something useful for our changing planet. He offers a philosopher's clear explanation of absolute versus relative, and of how the wilderness question fits into philosophy as a whole. He introduces the problem of translation: people speak of wilderness and mean different things, because they may be speaking from different cultures.

We need to encourage some kind of agreement, such as the International Union for the Conservation of Nature, the (IUCN) World Conservation Strategy, or the United Nations Environment Programme (UNEP)'s Agenda 21. Henberg demonstrates how analytic philosophy helps us understand the wilderness and shows the relevance of abstract theory to practice.

Lois Ann Lorentzen raises the question of women and the land—is wilderness a male idea, a by-product of the spirit of conquest and peakbagging? Initially she notes that feminist peasant movements in Central America have little use for the distancing term *wilderness*—the women involved are far too involved with the day-to-day use of the Earth to imagine it is a romantic, fearsome, or wonderful place apart from them. At the same time, Lorentzen speaks of her own sense of the wild while climbing in the Sierra Nevada, herself caught between the ambiguities of woman, nature, the North, and the South. There can be no simple identification of women with the Earth, and no simple rejection of the valid presence of the wild, either.

Douglas Buege approaches the Inuit of northern Canada with a sense of deep self-questioning. He is a traveler from the industrialized West, laden with presuppositions about what the North should stand for. He tells of prejudices of previous white voyageurs, and he is not afraid to admit his own. Caught somewhere between anthropology and reflective philosophy, he begins with the anomaly of his own presence in an alien culture, and proceeds to examine what responsible behavior ought to be in this foreign place, for both natives and visitors. Drawing on feminist theories, he outlines a new conception of responsible knowing for use in the North specifically, and in the wild in general. (His essay provides a sobering contrast with my own later contribution, which tends much more toward an easy romanticizing of the North as place for the wild.)

The Art of the Wild

The next group of essays takes off from the notion that wild places demand a wild culture, a different way of perceiving, writing about, and celebrating the extremes of the natural world. David Abram applies the phenomenology of Heidegger and Merleau-Ponty to propose a way to learn to feel the world space-time as one, to reconceive the world as a habitable wilderness known a whole new way, with our basic metaphysical questions of where and when exploded into something originary and transformative, meant to draw us in to the depths of the perceivable universe. He wants nothing less than to blow apart basic preconceptions of physics, insisting that space and time can be one and the same if we make ourselves as wild as the surrounding world. Here he is nearly echoing bards such as Guillaume de Machaut and T. S. Eliot, who, five hundred years apart, voiced the same yen to collapse time into itself: "In my end is my beginning, in my beginning is my end."

All sides of time are right before our eyes, but they are not the same. As French director Chris Marker said in the landmark film *La Jetée*, "The future is more protected than the past." This revelation of coordinates is part of a strategy that makes the wild more accessible to an enlightened humanity. And we need to think of time and space anew. Consider, for example, the words of a consultant to the U. S. Department of Energy, now planning for the far distant future of our nuclear waste under Yucca Mountain in Nevada: "Human beings have gotten pretty good at looking into deep space, but we are really no good at looking into deep time."[13]

Irene Klaver plays in the spirit of Abram's approach. She explores the phenomenon of the wild through the means of silence, interpolating fragments of experience and texts with the elusive search for the wolf, in and out of the American West. Sifting through memory, anecdote, and observation, she tries to pick out the wild in things by letting them speak for themselves. This is phenomenology in action, cutting through the abstruse terminology promoted by verbal tricksters such as Edmund Husserl and Martin Heidegger. There is no conclusion here but rather a juxtaposing of attributes: the silent and the wild, with the seldom-seen wolf as the living, leaping bridge between them. "The wild," writes poet Tomas Tranströmer, "does not have words."[14] We speak in spite of it.

My contribution is an attempt to make philosophy itself more like an ice floe, more lyric than analytic. Wittgenstein cryptically advised us that philosophy could only be written in poetry in the coming age, and cautioned that the kind of poetic thinking he aimed for was still not something he was writing himself. I submit that he should have looked out of doors, should have explored, should have responded to the land.

Our directions—north, south, east, west, and all degrees in between—are not locations on the fixed map of space and time, but are ideas, philosophical premises that guide the attempt to find sense of place. I choose the North here, mixing original reminiscences with those of fellow travelers, some who hunt for stories and others who make them up. Is it romanticizing? Buege and Cosgrove might think so. At least I want it to be *true* romance, accessible to all cultures, perhaps foreign to all as well, except those that can accept the crash of two icebergs as a semantic event.

Composer Murray Schafer is probably the foremost musical environmentalist, or muse-echologist, of our time. His book *The Tuning of the World*[15] describes the earth as a vast soundscape, listenable as huge musical composition or ultimate aural inspiration. Sounds, he tells us, are as endangered as other components of primal wildness. Woody Guthrie knew this as well:

> Music is a tone of voice—a tone of nature, a sound life uses to keep the living alive and call us back many times a day from the brinks of torture and the holes of superstition. Music is in all of the sounds of nature and there never was a sound that was not music—the splash of an alligator, the rain dripping on dry leaves, the whistle of a train, a long and lonesome train whistling down, a truck horn blowing at a street corner speaker—kids squawling along the streets—the silent wail of wind and sky caressing the breasts of the desert. Life is this sound and since creation has been a song and there is no real trick of creating words to set to music once you realize that the word is the music and the people are the song.[16]

Schafer's essay here is a précis of his fabulous composition *The Princess of the Stars*, a myth to be performed at dawn on a wilderness lake. This is what art in the wild might be: not a representation of nature, but a way into wild places that leaves space for nature, teaching us to listen, drawing cultures together with plural respect for the beyond-human cycles. The sun is a character, the birds sing their parts. We are all moved to tears as the stars cry themselves out and the day begins.

The Wild Revised

The final section presents direct challenges to the notion of wilderness, pointing out where it needs to go to survive into the twenty-first century as a guiding notion for ecophilosophy and practical environmentalism. First is Tom Wolf's engaging case study of predators in the Sangre de Cristo mountains of Colorado, part of his landmark book-length study currently in preparation. Here we learn of one unique place and its changing management status as wilderness. Maverick forester Arthur Carhart is contrasted to the better-known Aldo Leopold, revising notions on the best course of action for a place we have pledged to preserve in its pristine state.

Psychologist Robert Greenway has taken students into the wilderness for more than twenty-five years, and is now searching to articulate just what kind of good it does people to take them out of doors and away from their familiar worlds. Although the therapeutic value of hiking the backcountry has been assumed for decades, few have sat down and tried to state exactly what benefit such a trip provides. Greenway muses on this value here and concludes that an authentic journey into the wilds can be a healing experience because it may cut through the dualism enforced by our mundane culture.

Andrew Light throws a curve into our evolving history of nature. He argues that the city is the place where the classical sense of wilderness still holds: the sense of untamed, rampant danger. Meanwhile, wilderness as place of solace and preservation is a romantic idea, something that has gotten far too much inspirational press. What does this say about what we have done to the wild and the tame, to culture and to nature? Can he be this suspicious? Does wilderness not exist at all? Both classic and romantic kinds of wilderness are cultural constructs, and our notion of nature will always be in flux.

"Let them be left, O wildness and wet!" We want to leave it alone, but it is we who have put it there, with our relentless gaze and need to name things. "Ah, wilderness . . . ," says the mapmaker and scrawls the name over the white space, leaving some unknown for the would-be traveler to enjoy. Today we draw hotly debated lines that decry this place within wilderness, this other place without. Inside us, a piece of the wild always remains. Let the climb begin.

Notes

1. John McPhee's *Encounters with the Archdruid* (New York: Farrar, Straus, Giroux, 1969) is still one of the best-told tales about a conservationist and his natural enemies, pitting David Brower against adversaries Floyd Dominy of the Bureau of Reclamation and the Glen Canyon Dam, Charles Fraser of Hilton Head Island, and Charles Park of the U.S. Geological Survey.

2. J. Baird Callicott, "The Wilderness Idea Revisited: The Sustainable Development Alternative," *Environmental Professional* 13 (1991): 235-47.

3. Aldo Leopold, *Sand County Almanac* (New York: Ballantine Books, 1970 [1949]), 262.

4. Holmes Rolston, "The Wilderness Idea Reaffirmed," *Environmental Professional* 13 (1991): 370-77.

5. Ibid., 376.

6. Philip Booth, "Distances/Shallows/Deeps," *Ohio Review: Special Issue on Art and Nature* 49 (1993): 18. Emphasis in the original.

7. John Muir, "The Wild Parks and Forest Reservations of the West," *Atlantic Monthly* 81 (1898): 15.

8. Jack Turner, "The Abstract Wild," in *On Nature's Terms,* ed. Thomas J. Lyon and Peter Stine (College Station: Texas A&M University Press, 1992), 103.

9. Roderick Nash, *Wilderness and the American Mind,* 3rd ed. (New Haven: Yale University Press, 1967), 141.

10. Peter Matthiessen, *The Snow Leopard* (New York: Viking, 1977).

11. Roland Goetz, comment at the Fifth World Wilderness Congress, Tromsø, Norway, October 1993.

12. For more on this, see *Wisdom in the Open Air: The Norwegian Roots of Deep Ecology,* ed. Peter Reed and David Rothenberg (Minneapolis: University of Minnesota Press, 1993).

13. Kai Erikson, "Out of Sight, Out of Our Minds: A New Species of Trouble," *New York Times Magazine,* March 6, 1994, 63.

14. Tomas Tranströmer, "From March '79," trans. John F. Deane, in *Selected Poems 1954-1986,* ed. Robert Hass (New York: Ecco Press, 1987), 159.

15. R. Murray Schafer, *The Tuning of the World* (New York: Knopf, 1977).

16. Woody Guthrie, *Pastures of Plenty,* ed. Dave Marsh and Harold Leventhal (New York: Harper Perennial, 1990), 106.

1 . WHOSE WILD IDEA?

1 . Wise and Sustainable Uses: Revisioning Wilderness

R. Edward Grumbine

As the third millennium A.D. approaches, ideas and images of wilderness in North America appear to be evolving toward some as yet unknown configuration. Evidence of these changes may be found in the number of recently published books and articles that critically reexamine various facets of the relationship between humans and wild nature. Philosopher Max Oelschlaeger has provided a developmental history of the idea of wilderness from the Paleolithic to the present.[1] There has been a lively exchange in *The Environmental Professional* over the role of wilderness as a cultural ideal and conservation strategy.[2] Biologists Reed Noss and Hal Salwasser also have debated similar issues in the literature of conservation biology.[3] And botanists Arturo Gomez-Pompa and Andrea Kaus have weighed in with an attempt to "tame the wilderness myth" in the pages of *BioScience*.[4]

Dialogue over fundamental cultural matters does not take place in a vacuum. There are at least three reasons why wilderness is being reexamined today. The first reason derives from the pre-Darwinian roots of current Western conceptions of wilderness. In what Callicott labels "the received wilderness idea," people are seen as radically separated from nature and wilderness areas are considered to be pristine enclaves of nature, untainted by human handiwork, and operating in harmonious balance with the natural landscape within which they are embedded. This conventional image, familiar to both wilderness supporters and those who wish to develop wildlands, has been legally codified in the 1964 Wilderness Act: "A wilderness, in contrast with those areas where man and his works dominate the landscape, is hereby recognized as an area where the earth and its community of life are untrammeled by man, where man himself is a visitor who does not remain."

This people/nature duality informs both resource conservation, whose adherents believe that natural resources exist to be utilized for human benefit, and wilderness preservation, whose followers argue that a significant portion of landscapes should be protected in an undeveloped condition. The paradigmatic examples of these two camps have always been Gifford Pinchot and John Muir. (Though many observers consider Muir to be the father of preservation, it must be noted that his thinking on the value of wild nature traveled well beyond that of most preservationists toward a unification of people with nature.) The upshot of the radical Western split between people and nature is that *both* resource conservationists and wilderness preservationists, as long as they view nature as a collection of resources for humans, inhabit a world that categorically denies the full range of symbiotic relationships that may exist between people and wilderness. And, by focusing on nature as a fountain of inspiration or source of products, modern people have neglected the ecological theater and the evolutionary play that drives the dynamic, ever-changing patterns and processes of Earth.

The second reason why the idea of wilderness is being critically reexamined is that science is finally beginning to offer theoretical and empirical insights into the ecological and managerial implications of the people/nature dual*ity*. Conservation biology is providing compelling evidence as to why the image of conservation versus preservation has not served humans or nature well. Many species populations are losing their evolutionary viability, ecosystem functions (e.g., nutrient cycles, the water cycle, and patterns of growth and decay) are being fundamentally altered by human activities, and at the biosphere level, the effects of greenhouse gases on Earth's atmosphere will likely affect nature reserves and managed landscapes in ways detrimental to both resource protection and extraction.

Viewed in historical perspective, ecological dysfunction, once limited to particular species and specific locales, is now systemic from endangered gene pools to planetary climate. Accordingly, conservation biology is attempting to understand the dynamics of dysfunctional processes and advocate alternatives. Some environmentalists have seized upon the new field as providing irrefutable evidence that current efforts to preserve wilderness are grossly inadequate. The biodiversity crisis is challenging the fundamental logic of pristine wilderness set-asides surrounded by intensely managed multiple-use lands.

The third reason why wilderness is undergoing reevaluation today is grounded in emerging political alternatives to the resource conservation/ preservation dichotomy. Two conflicting positions, either of which would alter current wilderness policies if implemented, are jostling for attention.

One alternative to the status quo, being explored by a broad spectrum of both conservationists and preservationists, is *sustainable development*. Many claim that sustainable development offers a long-term antidote to the problem of humans destroying ecosystems more quickly than they can be renewed by ecological processes.

The second position, represented by the Wise Use Movement (WUM), has little broad-based political strength (as of yet). The WUM would like to expand the human hegemony over nature. WUM supporters "seek unrestricted access to all natural resources for [private] economic use, benefit, and profit."[5] This position, of course, is at odds with the sustainable development concept. It is also completely opposed to wilderness preservation in any form. But although the WUM may appear too radical to garner wide support, its potential political influence is not to be taken lightly.

The unresolved Western split between people and nature, the biodiversity crisis, and nascent alternatives to long-standing land management policies provide insights into why the idea of wilderness is undergoing intense scrutiny in the 1990s. Because these factors are interrelated and go down to cultural bedrock, they are difficult to untangle. As Oelschlaeger points out, any attempt to review the assumptions of modernity, in which the relationships between humans and nature are clearly central, is difficult because "through the lens of history human experience takes place entirely *outside* nature" (emphasis added).[6] The idea of history in the West has subsumed wilderness to the degree that we find ourselves living, as Aldo Leopold realized fifty years ago, in an ecological "world of wounds."[7]

For the 1990s, the upshot of this cultural inheritance can be summarized succinctly: Ecologically, we have erected a system of parks and wilderness areas that preserve scenic landscapes but show little regard for ecological integrity. Politically, we have assumed that such legal designations would be sufficient to protect wild nature. Philosophically, we have spent much of our energy on debates over intrinsic versus instrumental values in environmental ethics. Experientially, with more people having less access to wildlands as populations grow and urbanization continues apace, we have maintained, if not strengthened, the boundaries between wilderness and civilization.

Given the current rate and scale of ecological deterioration and the depth of our cultural predicaments, it is unfortunate that many of the writings reexamining wilderness are shot through with arguments that tend to obfuscate rather than illuminate many critical points. My aim in this paper is to reveal some of the conflicts that persist over the idea of wilderness and clarify areas that need attention. I also wish to provide notes toward the future of wilderness policy in North America as our ideas move away from viewing the world as a collection of resources toward an ecosystems view and beyond. I shall focus on two problems: the absence of a clear distinction between wilderness and wildness, and the role of wildness in any proposed revision of land management theory and practice.

Saving Wilderness or Retaining the Wild?

"I wish to speak a word for Nature," begins Henry David Thoreau in his essay "Walking," "for absolute freedom and wildness, . . . to regard man as an inhabitant, or part and parcel of Nature, rather than a member of society."[8] Thoreau uses the term *wildness* here, not *wilderness*, as is also the case later on in the essay where he states emphatically that "what I have been preparing to say is that in Wildness is the preservation of the world." Callicott suggests that Aldo Leopold, a century later, was also "concerned primarily . . . with integrating an optimal mix of wildness with human habitation and economic utilization of the land."[9]

Yet, if recent reexaminations of the wilderness idea provide any indication, we are still confused about exactly what Thoreau and Leopold were driving at, that wildness and wilderness are not equivalent in definition, meaning, or importance. Until we get clear on wildness and wilderness, we cannot be prepared to envision, let alone act upon, practical alternatives to the current ecological crisis.

In Oelschlaeger's view, humans have been erecting a boundary between themselves and nature since the advent of history. This "fence" has become increasingly rigid over the centuries; the existence of the people/nature dichotomy is one of the key assumptions of modernity. Civilization is both idealized and experienced as antithetical to wild nature. This assumption is not inconsequential—it allows most citizens of modern societies to inhabit a world that is, by any scientific reading of the facts, being destroyed by industrial imperialism. The people/nature split has led us to focus our efforts on preserving wilderness instead of protecting wildness.

The idea of wilderness is culturally relative, of course, a fact pointed out by several observers.[10] Many nonindustrial cultures have no word or concept for wilderness. And, despite all the lofty pronouncements of preservationists, wilderness is still viewed by most citizens of industrial societies as a resource for humans.

Wildness, on the other hand, as "the process and essence of nature," is the *source* of resources and of human existence.[11] It is the generative framework within which all beings inhabit Earth. Wildness escapes easy definition. Though it permeates life, it is easier to indicate its presence than to pin it down conclusively. While wilderness in Western culture is most often a place, wildness is the force behind places, "the overarching reality that transcends all our plans and creations."[12] A biologist might study natural selection and adaptive mechanisms in the field but the process of evolution is wild. Wildness belies the possibility of the mechanistic worldview of industrial civilization triumphing over nature. Even nuclear winter, the most dismal environmental catastrophe imaginable, though it would cause great suffering and radically diminish life, could not totally extinguish the regenerative force of wildness.

Humans, too, are wild. Yet our individual personhood is embedded within and fully informed by events of the "external" world. Our fundamental wildness presents itself continually—we breathe without effort, we exchange matter and energy with the world, we are aroused by sudden sound and movement, we wonder at lightning storms and cloud patterns. Wildness in humans might be characterized as the self-regulating aspects of the *body* interacting with the unconscious depths of the *mind* with each of these in constant contact with *environment.*

Both Thoreau and Leopold saw wilderness as a cultural construct flowing from and dependent upon wildness. In "Walking," Thoreau was careful to distinguish between inhabiting nature and being a member of society. For Thoreau, inhabiting (though he did not provide an explicit definition) was essential for maintaining both freedom *and* wildness in human culture. The two were inevitably linked for, if in wildness the world was to be preserved, then the loss of wildness would lead to certain ruin. Thoreau was probably the first person born of modernity to recognize the grave consequences of industrial civilization's project to dominate nature and contain wildness. Gary Snyder and Neil Evernden remind us again of the distinction between wildness and wilderness.[13] Snyder both clarifies and critiques Thoreau's bold assertions when he says that "wildness is not just the preservation of the world, it *is* the world."

Wilderness areas, of course, may allow wild nature to live and breathe to the degree that they are not subject to human control. And they are important (to humans) to the extent that they allow direct contact with wildness, which may result in experiences that transcend the culturally relative categories of modern existence. Wilderness and wildness intersect where a river, mountain, bear, or beeplant sparks an awareness in us that helps to break down the fence between people and nature, where value and valuer (or, to the use the axiological categories of contemporary environmental ethics, instrumental and intrinsic values) appear as limited constructs.

Still, the importance of protecting wildness goes deeper than the potential for healing the people/nature split. This dichotomy itself springs from a fundamental paradox of human existence: the distinction between self and other. Oelschlaeger argues that while "the Paleolithic mind did not distinguish the human enterprise from the natural world, ... it did wonder at the miracle of existence and created an elaborate hunting mythology to account for reality."[14] If this is true, then consciousness of the other is not simply a phenomenon of history, but is part and parcel of human experience at least as far back as the formation of (proto-) culture. Wildness bears on the distinction between self and other, not in terms of erasing differences, but in the recognition of organic connections. The resolution of the paradox is found not in denying the distinctions between humans and other species, wolves and invertebrates, or any members of Earth's community of life, but in what we decide to make of the differences.

Philosopher Tom Birch is explicit concerning the usual Western position on the paradox of the other: "[m]ainstream Western culture ... presupposes that opposition is fundamentally conflictive, rather than complementary ... or ecosystemic."[15] Taking conflict with otherness as our standard, we have narrowed our relationship with wildness to a combination of two approaches: (1) gaining power by dominating the world through objectifying the diversity of life and reducing it to resources for human consumption; and (2) simply obliterating nature on a large scale and replacing it with developments (cities, factory farms, parking lots) of various kinds.

Both of these responses reinforce the Western concept of people being separated from nature and somehow outside the forces of evolution. But if evolutionary pathways may be modified by culture, but not ultimately overcome by cultural adaptations, then industrial societies have a problem. To paraphrase writer Barbara Allen, we haven't lost our relationship with wild nature, we have simply invented it in terms that do not allow us to erect a sustainable, cooperative relationship with it.[16]

In adapting ourselves to fit the "rules" of wildness, we may choose between fear of, or solidarity with, wild nature. This choice, however, is not cut-and-dried. It is always bound up with ongoing human predicaments such as the need for each individual within a society to come to terms in his or her personal development with the broad answer to the self/other paradox that the culture has already imposed. New ways of integrating people and wildness must first grapple with the old answers to this paradox, and Western culture has a well-established track record in the pursuit of domination. Each individual must furthermore define through personal experience just how widely the boundaries of the self may be extended toward the other. Answers to these issues are always provisional and more or less in process. Evernden suggests that "perhaps even wildness is an inadequate term, for that essential core of otherness is inevitably nameless, and as such cannot be . . . made part of the domain of human willing."[17]

To imply, however, as Callicott does that the concept of the other *reinforces* the people/nature split is to ignore this fundamental paradox that is, has been, and will continue to be part of the human condition. And, to state, as Callicott has done recently, that "the ubiquity of man and his works has made the illusion of nature as Other all but impossible to maintain"[18] is to confuse the eternal *presence* of otherness (shall I say wildness?) with the Western attempt to subdue it.

The modern concepts of wilderness, wildness, and self/other may indeed be evolving. But old worldviews do not dissipate quickly, and behaviors that put into practice new ways of being take even longer to become established as we feel our way from conflict toward complementarity. If a revised idea of wilderness does have a role in a sustainable future (and I believe emphatically that it does), then we must hook it to protecting wildness. And if humans and their cultures also are fundamental expressions of wildness, then we must allow for wildness (and wilderness) to flourish in any conception of sustainable development.

What Does Sustainable Development Sustain?

Many of Western culture's difficulties with allowing wildness to exist are revealed in the arguments of writers who would substitute for the idea of wilderness the concept of sustainable development. The debate between two of the principals in the discussion, philosophers J. Baird Callicott and Holmes Rolston, is grounded in contrasting views of the place of people in nature.

Rolston affirms a radical discontinuity between nature and culture, wilderness, and civilization: "Humans now superimpose cultures on the wild nature out of which they *once emerged*."[19] He believes that human impositions upon nature, because they are the product of "deliberated human agency," are inevitably "artificial, unnatural." For him, in the wilderness debate culture "is a postevolutionary phase of our planetary history."[20] There could not be a more concise statement of separation between people and nature. According to Rolston humans, by virtue of their self-reflecting capacity and ability to intervene in the order of things, have today reached some kind of escape velocity and exited the evolutionary loop. He recognizes a historical progression of culture evolving "out of nature" even while he champions respect for "our wild origins and our wild neighbors on this home planet."[21] But his support for wildness is problematic. In the context of modernity's attempt to dominate nature, otherness, to those who consider people to be "postevolutionary," likely remains an adversary, a competitor, a force against human projects. Given this, it is difficult to see how otherness can be accommodated, respected, or loved. If humans cannot recognize wildness in themselves or their cultures, it is doubtful that they will be able to respect wildness in either wild places or wild nonhuman beings.

At first glance, Callicott appears to offer an alternative to Rolston that breaks down the people/nature fence and sets us firmly on a path toward re-visioning wilderness. Callicott does not place people outside of nature. He "follow[s] Darwin in thinking that human culture is continuous with primate and mammalian proto-culture and that, no matter how hypertrophic it may lately have become, contemporary human civilization remains embedded in nature."[22] Reviewing the Western idea of wilderness, Callicott observes correctly that it perpetuates the human/nature dichotomy, that it is ethnocentric to the degree that it ignores the historical presence of people living in "pristine" ecosystems worldwide, and that it paints a static picture of nature as if, for example, the Yosemite Sierra in California would perpetually remain as it was when the area was designated a national park in 1890. As an approach to conserving wildness, Callicott would scrap the outmoded Western concept of wilderness and replace it with a "postmodern, technologically-sophisticated, scientifically-informed" sustainable development that would reintegrate people with nature by limiting human enterprises to those that do not "compromise ecological integrity seriously."[23] As a paradigm for sustainable development, Callicott holds up Aldo Leopold's concept of land health ("Conservation is a state of harmony between man and land")[24] and suggests that it presents an alternative to the use versus preserve status quo.

Callicott's is a genuine attempt to move beyond the either/or nature of the wilderness debate. But he has not taken us far enough in the search for sustainable relationships between people and wildlands. First, he does not consider the fundamental self/other paradox as part of the basic conditions of life on Earth for humans. Under the best of circumstances people will always be engaged in learning from and adjusting to the ecosystems with which they are in partnership. To state this in terms of Leopold's ideal of conservation, harmony between people and nature is not so much a balanced state of grace as it is a dynamic complementarity that must be continuously renegotiated as individuals, cultures, and ecosystems evolve. From this perspective, the issue is not whether we can break down completely the wilderness/civilization dualism but whether we can reduce conflict and increase cooperation. With this distinction in mind, we can begin to comprehend some of the limits of Callicott's arguments for a sustainable development alternative to wilderness.

To support his charge of ethnocentrism, Callicott cites new evidence that humans have lived for thousands of years in what the Euro-American West has always considered untouched wilderness. A corollary of this is the degree to which primal peoples altered ecosystems before the arrival of Euro-Americans. From an ecological perspective, these issues are primarily questions of the rate and scale of human impact on the land. Before modern technology, the transformative power of humans was many times less than it is today. The fact that biologists and anthropologists are "discovering" that the Amazon Basin was "densely" populated and modified by humans says less about the state of wildness or land health of the region than it does about the persistence of modernity in denying the value (and even existence) of anything not part of the Euro-American image of wilderness and civilization. Is Guatemala's Tikal National Park, part of the vast, sparsely settled Petén region of tropical forest, less healthy today even if we know that one thousand years ago much of the area was clear-cut and cultivated by the Maya? Such information may prove Callicott's charge of ethnocentrism. However, the important point is not whether people lived (or live) in wild places, but *how* they lived and continue to dwell in wildlands while accommodating wildness, and what we might learn today about land health from considering their lifeways.

Callicott offers several suggestions for integrating a Third World approach to conservation with efforts in the First World. All his suggestions assume, however, that Third World peoples must be brought into the framework of the global, industrial economy. Extractive reserves and ecotourism, both currently in vogue with international conservation groups, will likely only continue Third World dependence on the industrial models of the First World. Such policies might teach forest peoples about the vagaries of global prices, but will do little to educate the citizens of industrial societies with the wisdom derived from lives lived more closely following Leopold's conservation ideal. If there is to exist a sustainable development that enriches the wilderness idea, there must be choices beyond either trying to integrate people into the global cash economy or continuing the destruction of wild ecosystems. Neither of these genuinely furthers the quest for a working definition of land health à la Leopold or help reduce conflict between people and nature. A third path lies in stimulating nonindustrial local economies with local products for local use while simultaneously reducing both the rate and scale of the First World nations' consumption of resources. This is much easier said than done, but we do not want to perpetuate past problems as we envision new sustainable approaches.

The main problem regarding wild places is how to increase their biological integrity while constraining human development. If sustainable development is to succeed, there must be limits on how much habitat humans appropriate for themselves to the detriment of other living beings. This is problematic—the United States, a country that is very rich and relatively sparsely populated, has been able to formally protect only about 5 percent of its land from development. By most ecological accounts, this is nowhere near enough. For example, according to the best current information, protection of old-growth Douglas fir ecosystems and their host of dependent species in the Pacific Northwest will likely require a decrease in the region's timber cut by more than 80 percent from recent levels.[25] Columbia River salmon populations have plummeted from 10 to 16 million fish in 1850 to about 2.5 million today with 75 percent of current fish from hatchery stock.[26] At the national level, 59 percent of all species listed under the Endangered Species Act are either declining in population, extinct, or of unknown status.[27] If one looks at the forest landscapes of North America and includes ecosystem processes such as wildfire in one's definition of land health (as Leopold surely would have), the disturbing conclusion is that, because the U.S. Forest Service has actively suppressed fire for almost a century, hundreds of millions of acres need a good clean burn today. Worldwide, three-quarters of all bird species are declining in population or threatened with extinction, while the entire class of amphibians is losing ground.[28] And Soulé estimates that nonhuman primates, large carnivores, and most of the hoofed animals may "all but disappear" within the next century.[29] Assuming these data provide a rough estimate of how nonsustainable industrial culture has become, it may well be that conservation biologist Reed Noss's suggestion that 50 percent of the lower forty-eight states be protected in reserves "where humans do not dominate" does not go far enough.[30]

Given his knowledge of the biodiversity crisis, Callicott's disagreement with Noss's working proposal on the grounds that it is implicitly impractical, unreasonable, and misanthropic, borders on the disingenuous. Callicott has neglected to make clear the *limits* of sustainable development. To define these limits, we must first ask three questions: What are we trying to sustain?

What are we attempting to develop? Who will benefit from these actions? If, following Callicott, we are trying to sustain biodiversity and ecosystem health, then we must explicitly recognize that a sustainable development alternative can succeed only if it is grounded upon an adequate system of ecological reserves. Callicott advocates "big wilderness," a national system of ecologically defined reserves, cessation of old-growth logging, and elimination of livestock from western U.S. rangelands. All too often, however, his language betrays a tendency to portray humans as agents of control ("The past affords paradigms aplenty of an active, transformative, managerial relationship of people to nature," [31] and "a new generation of postindustrial technologies may make it possible for us to pursue many of our economic activities without compromising ecosystem health"[32]). Callicott even offers Aldo Leopold as the head of an "ecological farm family [that] *actively manages its wild lands*" (emphasis added).[33] And, if both people and nature are to be the beneficiaries of the sustainable development alternative to wilderness, as Callicott would prefer, asking "Can we succeed as a global technological society in enriching the environment as we enrich ourselves?" poses problems for those who do not wish to be linked to such a new world order. Given the thrust of modernity and the depth of the biodiversity crisis, a "global technological society" is hardly compatible with "ecological exigencies." I believe that such language, lacking humility and emphasizing management over restraint, is dangerous and as problematic as Rolston's denial of the fundamental wildness of human nature. There is a great deal of hubris here and very little of the sense of limits that will be required of humans in any transition toward sustainability.

Perpetuating Conflict: The Wise Use Movement (WUM)

Sustainable development is not the only general environmental policy option being discussed today. While environmentalists attempt to institute international debt-for-nature swaps, ecotourism, and extractive reserves, and scholars debate wilderness in academic journals, the WUM has grown quickly from a handful of individuals to a coalition of some 250 loose-knit groups with a common agenda.

The WUM, as the name implies, does not focus exclusively on wilderness. But the movement's vision of working relations between people and nature is so narrowly construed as to be anathema for wildness in particular and sustainable development in general. The WUM grew out of the Sagebrush Rebellion of the early 1980s. Its philosophical lineage in the United States can be traced back to what historian Craig Allin has called the "economics of superabundance," where resources are always unlimited and only labor is lacking.[34] The movement is a caricature of the "wise use" of Gifford Pinchot, who first brought the concept to American conservation.

Most WUM supporters do not believe in either public lands or federal land management and seek to replace these with a no-holds-barred utilization of resources for private profit. The movement's manifesto, *The Wise Use Agenda*, includes the following among twenty-five goals: open access to mineral and oil resources in all wilderness areas, national parks, and wildlife refuges; logging and replanting of all U.S. old-growth forests; amendment of the Endangered Species Act to exclude "non-adaptive species" and "species lacking the vigor to spread in range"; and elimination of wetlands development restrictions.[35]

Several charismatic leaders have worked together to create the organizational vision for the WUM. Ron Arnold, executive vice president of the Center for the Defense of Free Enterprise, a group that espouses an extreme form of free market economy, is the publisher of the *Agenda* and a key coordinator. Charles Cushman, president of the Multiple Use Land Alliance representing landowning inholders within national parks and national forests, is another important actor. People for the West!, the main lobbying group supporting the 1872 Mining Law, also figures prominently. These three organizations are accompanied by an extraordinary diversity of member groups, including farmers, loggers, western water interests, cattlemen, off-road vehicle users, beach developers, miners, and more. Most of these groups appear to be small and grassroots in political scope in contrast to the national organizations represented by Arnold, Cushman, and others. This cleavage in the WUM has yet to manifest itself in the movement's political efficacy.

The financial clout of the WUM movement is substantial. Extractive industries (mining corporations, timber companies), trade associations (American Farm Bureau, National Cattlemen's Association), and off-road vehicle manufacturers (Honda, Kawasaki) have provided most of the funding up to the present. Direct mail contributions from the growing membership are just beginning to be exploited. Alan Gottlieb, another leader of the Center for the Defense of Free Enterprise, noted in 1992, "In the past five years we've raised $3 million for Wise Use issues, and $1 million of that came in the last year. The potential is way, way greater. We can reach five million households rather quickly."[36]

The message that the WUM is attempting to spread is based on the two interrelated fears of losing access to the resources upon which the extractive economy depends, and losing jobs. Strategies adopted to carry this message to members have, at times, approached what might be termed "frenzy politics." For example, when organizing protests against a National Park Service ecosystem-based plan for the Greater Yellowstone ecosystem, WUM leaders bused hundreds of ranchers, loggers, and miners to public hearings and then, prior to entering the meetings, agitated the crowd by misrepresenting the plan as a threat to jobs and property rights. Park superintendents were labeled "Nazis" and environmentalists, "Communists." WUM organizers working on other issues have shown a similar propensity toward employing falsehoods and exaggerations in media campaigns that are based in conflict and alarmism.

The WUM appears to have adopted some of the tactics of the Earth First! movement, such as widespread media manipulation, inflammatory rhetoric, and appeals to the "grass roots." (People For the West! even employs an exclamation point in its name.) WUM tactics, however, are bereft of the humor, intelligence, and ecological vision of Earth First! Regardless of tactical similarities, the WUM differs in two important ways from radical environmentalism, both of which bode ill for wildness and sustainability—first, its ability to generate $1 million in a year from its supporters, and second, its ability to develop a mailing list of five million households.

Fear and loathing are the operative terms to describe the WUM's tactics, but the coalition is still a movement in the making. While the leaders, goals, membership, and strategies are relatively well known, it is not yet evident how influential the WUM will become in environmental policymaking. For the present, the WUM is focusing on several issues that relate directly to wildlands protection and sustainable development. These include opening all parks and wilderness areas to developers and motorized vehicle users; revising administrative, county-level land use codes throughout the western United States in an attempt to weaken on-the-ground application of federal environmental laws (WUM leaders claim that 25 percent of all counties in the West are involved);[37] opposing reauthorization of the Endangered Species Act; actively fighting any new federal legislation protecting old-growth forests in the Pacific Northwest; and opposing any reform of the 1872 Mining Law. Arnold even claims that the WUM's political clout extends internationally: "Bush didn't sign the biodiversity treaty [at the United Nations Conference on Environment and Development in Rio] because of us, quite frankly. We put on pressure. There was a lot of mail."[38] Whether or not this claim is true, numerous western congressional representatives actively endorse *The Wise Use Agenda*. Former Secretary of Defense Dick Cheney sits on the board of the Center for the Defense of Free Enterprise. Given the environmental track record of various administrations since 1980, it is easy to see how many of the goals of the WUM dovetail closely with conservative political agendas.

The WUM, like the concept of sustainable development, appears to be a response to life in the 1990s where "environmental protection" is no longer at the margin of economic activity. It might appear ironic that a movement bent on returning to the laissez-faire economics of superabundance approach to nature of nineteenth-century America would spring to life in an era of obvious limits. But, given the historical trend in the United States of increasing land use regulations (including wilderness designation and management) in response to declining amounts of resources such as roadless areas, productive grazing lands, uncut forests, and economically recoverable oil, gas, and mineral deposits, the rise of the WUM represents a last-ditch attempt by the most radically utilitarian members of society to maintain their nonsustainable lifeways. While Callicott and others, however imperfectly, endeavor to conceive of a sustainable human partnership with nature nested within ecosystems, WUM supporters comprehend nature as embedded within a free market economy based solely on the generation of private profits. Both sustainable development and the WUM do share common ground—each emphasizes the human *use* of nature. For Callicott, sustainable development is proposed as a means for protecting biodiversity and ecosystem health. For the WUM, whose members see no value in wildness and little human relationship with nature beyond the pecuniary, use is the beginning and the end of the story. The WUM is certain that success lies in perpetuating the old models of conflict that have strengthened the fence between people and nature. As a policy option, to the degree to which cooperative behavior must figure in protecting wildness and producing sustainability as I have argued here, the WUM has nothing to offer the future. But for those exploring the relationship between wildness and sustainable development, use is at once more complex and more difficult to frame into policies that would help to deemphasize the people/nature dichotomy.

Wildness Integrated with Sustainable Landscapes

Aldo Leopold, according to philosopher Bryan Norton, envisioned conservation "as one culture's search for a workable, adaptive approach to living with the land," recognizing that "if a society's practices are not adaptive, . . . the culture will fail."[39] Though it may appear to most observers that the

WUM supports adaptations contrary to Leopold's sense of fitness, it does not necessarily follow that revisioning wilderness along more sympathetic lines will provide a better fit.

Part of the problem lies in the fact that, outside of the "received" idea of wilderness, there are few if any models in Western culture that value wild places or speak to how wildness may positively influence humans. It has not been a dominant part of the Euro-American tradition to support wildness *or* economies of sustainability. In making a bid to substitute sustainable development for wilderness, it is as if we are attempting to find our way in deep woods without a map, or to build a bridge to the future by fitting square pegs into round holes. There are numerous ways to get lost or have our construction attempts fail. Rolston, as we have seen, for all his support of wilderness, disregards the bonds connecting wild nature with culture. Callicott, for his part, places both people and nature in a wild nexus but, focusing on "active," "transformative," and "managerial" concepts of sustainability that might even "improve" nature as well as conserve it, fails to invoke meaningful limits on human projects.

If we are not careful in our attempts at redefining the people/nature boundary to bring human uses of nature into a role supportive of wildness, wilderness, once thought to be part of *the solution,* becomes *the problem.* For example, Gomez-Pompa and Kaus, after reviewing the Western concept of wilderness and finding the same ethnocentric problems that Callicott points out, state that "a belief in an untouched and untouchable wilderness has permeated global policies and politics in resource management . . . causing serious environmental problems."[40] Yet it is difficult to imagine wilderness protection as contributing to environmental problems to the same degree as the development projects of industrial capitalism.

Since the Western idea of wilderness is "mostly an urban perception, the view of people who are far removed from natural environments they depend upon for raw resources,"[41] Gomez-Pompa and Kaus propose that we tame the wilderness myth by paying attention to rural people's experience of being at home in nature, where conservation is part of a lifeway and perception of working with ecosystems. These authors have touched upon a key to resolving the dilemmas of accommodating human use while sustaining wildness. But they place too much weight on the distinction between urban and rural values. One need only consider the WUM to acknowledge that there exists a wide range of rural attitudes toward conservation and wildness: Many rural people in the United States who derive their living directly from ecosystems are the same ranchers, loggers, and miners who form the grassroots constituency of the WUM.

A distinction more appropriate to the protection of wildness would be the one biologist Ray Dasmann has drawn between *biosphere people,* those who draw support from planetwide resources primarily through industrial fossil fuel economies, and *ecosystem people,* who get their living primarily within the constraints of local ecosystems.[42] The driving engine behind biosphere values, of course, is the Western image of people fenced off from nature, and the historical response of biosphere people to ecosystem people has been similar to that of biosphere people to wildness: domination. The value of Gomez-Pompa and Kaus's work is that they invite us to challenge this predominant view by reminding us that "throughout the world, communally held resources have been managed and conserved by diverse human societies via cultural mechanisms that attach symbolic and social significance to land and resources beyond their immediate extractive values."[43] But what lies "beyond" extractive value? By what specific "cultural mechanisms" may people today experience what these values entail? Does wildness have a role to play here?

Answers to these questions may be found by attempting to discover what values, if any, limit ecosystem people's use of wild nature. Callicott characterized the relationship between humans and large carnivores in North America prior to Euro-American contact as one of "mutual tolerance."[44] Similarly, Rolston declares that "the Indians did not need or achieve" the idea that ecosystems might be so respected that people would only visit and not remain.[45] Yet, these statements do not capture the importance of what today we might call the sacred and how such beliefs might limit human behavior in ecosystem cultures. A perusal of the anthropological literature on native people's relationship with grizzly bears, for example, suggests a far more meaningful relationship than mere tolerance. Paul Shepard and Barry Sanders report that the Bear Mother Myth, which "may be the most persistent and widely-told tale" in the Northern Hemisphere, tells of both identity and distinction between the species and outlines appropriate ceremonies to honor bears.[46] Complex rites honoring bears also are described by Irving Hallowell, David Rockwell, and Richard Nelson. Nelson considers a major tenet of Koyukon ideology to be the fact that "humans and natural entities are involved in a constant spiritual interchange that profoundly affects [i.e., limits] human behavior."[47]

Contrary to Rolston's suggestion, indigenous peoples often have developed relations with wild places in which only a limited human presence is allowed under certain circumstances. Snyder's account of a visit to a sacred place in the Australian outback in the company of aboriginal people of the Pintubi tribe proves that such practices still exist today, however tenuously.[48] My experience with the Mopan Maya of southern Belize has shown me that the Mopan do not travel to "wild" places indiscriminately and do not enter the forest without some modest ritual preparation.

The lesson here for biosphere people is that many ecosystem people are involved in relationships with their dwelling places that expand Dasmann's original concept of economically based resource behavior to include a spiritually based sense of enoughness that often limits their appropriation of what we would term "resources." Much evidence worldwide over the last several thousand years points toward lifeways based in respect and reciprocity that illuminate Leopold's original sense of conservation.

This is not to deny that hunter-gatherers and Neolithic agriculturalists caused extinctions at times. It is abundantly clear that people, as they migrated into new areas of the planet during the past 50,000 years, seem to have killed off numerous species. There are few if any reasons to suspect that there existed an ongoing harmonious balance between people and nature throughout prehistory any more than we can find evidence of static, unchanging ecosystems today. Humans of the past were not exempt from the self/other paradox, and it appears that we requires a period of adjustment as we settle in to new habitat before we can achieve any possibility of sustainable ecosystem behavior.

There are two distinctions, however, between human-caused extinctions of the past and those of the present. First, past extinctions generally resulted from the overhunting of particularly vulnerable (e.g., large or flightless) species, while those of today are due primarily to habitat destruction, resulting in an unprecedented rate and scale of human-caused extinctions. Second, there exist today accessible models of "postmigration" human behavior (ecosystem peoples) that seem to warrant the description "sustainable" from which biosphere people can learn.

Ecosystem behavior, whether supported in general terms by Leopold's land ethic or detailed specifically from anthropological accounts, depends upon a diminished fence between people and wildness. For, as Tom Birch points out, "a system of domination cannot grant full equality to *all* the dominated without self-destructing."[49] And, though profound cultural change is beyond the ken of environmental policymaking per se, policies subscribed to today may set the stage for changes over the longer term.

The Protection of Wildness Today

How do we begin to move from preserving wilderness to protecting wildness? How do we start to embrace ecosystem lifeways and decrease our dependence on biosphere lifestyles? What follows are notes toward answers to these questions for North Americans.

For the present, the first step in any such strategy is to *continue to focus on increasing the size and number of the protected areas we know as wilderness.* These lands are the last remnants of wild diversity and are faced with imminent development. But protection of biological diversity should be emphasized over preservation of scenic lands and recreational opportunities. Given the politics of wildlands protection in the United States, the best hope for success is to ground such an approach in science. A conservation biology-based platform would include (1) habitat protection for viable populations of all native species; (2) areas sized large enough to encompass natural disturbance regimes; (3) a management timeline that allows for the continuing evolution of species and ecosystems; and (4) human use integrated into the system of protected areas that would provide for *Homo sapiens* within the foregoing constraints.[50] Though the creation of such a system will be difficult, it is hard to justify anything less if sustainability *for all* is to be sought. Because most of the biodiversity and roadless, undeveloped areas are found on federal lands, this strategy must first focus on the public domain. But given the lessons learned from the use-versus-preservation approach to land management, it is obvious that this cannot remain the case for long. State and private lands must soon be brought into biodiversity protection partnerships.

Another step in the evolution toward an ecosystems partnership between culture and nature would be the implementation of ecosystem management within the landscape of use (multiple-use federal lands and all private lands). The four-planked conservation biology platform would apply to these lands as well. Many of Callicott's suggestions concerning the efficient use of nature with an accent on appropriate alternative technologies also could be applied here. This too is where various green alternatives to industrial culture could provide direction. And it is here where the language and ideology of sustainable development might metamorphose into that of sustainable landscapes. Landscape is a more appropriate image for sustainability than development for several reasons. It removes the focus on human projects implicit in "development" while also describing a place that provides room for all species. It includes human use without excluding nonhuman beings and their needs. And it may be made more specific as diverse peoples in many different places begin to discover what it means to live within the constraints of local ecological conditions. To paraphrase the writer Wendell Berry, our comprehension of sustainable landscapes will become more specific to the degree to which we begin to live fully in them.

The ultimate purpose of protecting wildness is not to preserve nature or improve upon it, but to learn a sense of limits from it and to model culture after it. The hope of protecting large wildlands with ecosystem management is that this strategy would slow the rate of the biodiversity crisis while also providing a model that could feed the nature/culture system *both* ways—sustaining wildness at the core of protected lands and at the center of human communities. The promise of this strategy is that as people begin to gain direct experience with ecosystems by working to protect biodiversity, wildness may explicitly become part of culture. It is just this sense of wildness that Thoreau was aiming for when he declaimed about "the preservation of the world."

Future Wild

Learning to live with nature in the near term will require much more than integrating landscape use with landscape protection. What Salwasser considers to be a "fact," that "more humans are going to demand more resources from remaining wildlands," must somehow be transformed into more humans demanding fewer resources through a combination of reduced consumption by biosphere people, greater technological efficiency, and an equitable redistribution of ecological goods and services.[51] Over time this must change into fewer humans demanding less. This prescription counters the very underpinnings of the political worldview of modernity and I do not mean to suggest that the transition will be easy if it occurs at all. I do know, however, that we must proceed as best we can and that time is running short.

Much help during this difficult transition could be obtained from two sources. First, more citizens must demand action on environmental issues from their political representatives. A rise in grassroots organizing on both sides of the fence (witness the plethora of regional environmental groups and the WUM) is much in evidence, but most citizens have not yet become involved. A second source of support must come from parents who would provide their children with direct contact with wild nature. With more than 75 percent of the U.S. population now living in urban concentrations removed from wildlands, this is problematic. Given the long-term aspects of the transition to sustainability, however, it is imperative that young children be connected with wildness. Without direct experience with wild plants and animals, mountains, deserts, and rivers, it seems unlikely that children will value nature as they grow older. Paul Shepard believes that the crucial stages of childhood development must be partially "enacted within [wild] ecosystems" and that the destruction of ecological diversity in both places and cultures has had profound negative affects on human psychology.[52]

This brings us to the role of symbolic behaviors that might encourage humans to integrate wildness with culture. After studying forest management policies and practices throughout Southeast Asia, M. G. Chandrakanth and Jeff Romm conclude that secular prescriptions do not capture all the important interests and concerns of people. Recognizing similar trends throughout the world, these authors suggest that debates over future policies "will not likely be resolved until [policies] are explained and treated in religious as well as in legal, economic, and ecological terms."[53] This points us once again toward a critical distinction between biosphere and ecosystem peoples. What Gomez-Pompa and Kaus described as "cultural mechanisms that attach symbolic and social significance to land" are rooted in experience of the sacred that, mediated through a cooperative approach to self/other, may lead directly to limits on behavior.[54] Put in terms of the biodiversity crisis, legal reform (creation of an Endangered Ecosystem Act), economic changes (inclusion of externalities in cost/benefit computations), scientific shifts (advancement in ecosystem management), and cultural revision of the Western idea of wilderness will not likely be sufficient to carry us very far beyond the threshold of sustainability.

For the long term, what we need are cultural practices that resacralize the world. Philosopher Michael Zimmerman characterizes these as ceremonies that would allow Western people to incorporate wildness into the definition of "civilized" by way of social rituals that would enable us to "solidify . . . identification with the forest families—trees and squirrels, deer and birds— and that would . . . simultaneously initiate [us] into a human family which had appropriate respect for and relationship with the other families of the forest."[55] This is what supporters of the deep ecology movement uphold as ecocentrism, transpersonal ecology, or an ecological approach to being in the world.[56] Such approaches seek identification with nonhuman beings and a wonderfully expansive boundary between self and other, where, in Snyder's provocative descriptions, "the flora and fauna and landforms *are part of culture*" and culture itself is "a nourishing habitat."[57]

Lacking this identification today, it is almost inconceivable to imagine how industrial culture might be so transformed. My comprehension of sustainability gives me hope that the details will be discovered in the living. For now, we must concern ourselves with charting the course so that others may sail the ship and have the chance for a successful voyage.

The tasks required by the quest for sustaining wildness may in time bring humans to large-scale ecological restoration, returning wolves and grizzlies to California, Quebec, and Mexico while linking wildlands across the continent. If so, we will have become a different people. By 2090, humans will have burned almost all of the planet's currently known reserves of fossil fuels. Young second-growth forests throughout North America in the 1990s will have become part of an established pattern of ancient forests. With the wounds perpetrated by industrial cultural healing, the fence between people and nature, self and other, will be partially torn down, blurred, indistinct at the margins. This vision of a sustainable future inspires hope. It renews commitment to what most certainly will be decades of difficult work. In time, however, we may finally begin to experience what it means to be partner with, rather than ruler over, nature. People, too, might become native— and wild—again.

Notes

1. Max Oelschlaeger, *The Idea of Wilderness* (New Haven: Yale University Press, 1991).

2. Reed Noss, "Wilderness Recovery: Thinking Big in Restoration Ecology," *Environmental Professional* 13 (1991): 225–34; J. Baird Callicott, "The Wilderness Idea Revisited: The Sustainable Development Alternative," *Environmental Professional* 13 (1991): 235–47, and "That Good Old-Time Wilderness Religion," *Environmental Professional* 13 (1991): 378-99; Holmes Rolston, "The Wilderness Idea Reaffirmed," *Environmental Professional* 13 (1991): 370–77.

3. Noss, "Wilderness Recovery," and Hal Salwasser, "Sustainability as a Conservation Paradigm," *Conservation Biology* 4 (1990): 213–16.

4. Arturo Gomez-Pompa and Andrea Kaus, "Taming the Wilderness Myth," *BioScience* 42 (1992): 271–79.

5. D. Callahan, *The Wise Use Movement*, W. Alton Jones Foundation, Washington, D.C., 1992 (available from Cascade Holistic Economic Consultants, 3758 S.E. Milwaukee, Portland, OR 97202), 2.

6. Oelschlaeger, "Idea of Wilderness," 7.

7. Aldo Leopold, *A Sand County Almanac* (New York: Oxford University Press, 1949).

8. Henry David Thoreau, *Excursions and Poems*, vol. 5 of *The Writings of Henry David Thoreau* (Boston: Houghton Mifflin, 1906), 205.

9. Callicott, "Wilderness Idea Revisited," 238.

10. Ibid.; also Oelschlaeger, "Idea of Wilderness"; Gomez-Pompa and Kaus, "Taming the Wilderness Myth."

11. Gary Snyder, *The Practice of the Wild* (San Francisco: North Point Press, 1990), 5.

12. Tom Lyon, "The Saving Wildness," *Wild Earth* 2 (1992): 3.

13. Snyder, *Practice of the Wild*, and Neil Evernden, *The Social Creation of Nature* (Baltimore: Johns Hopkins University Press, 1992).

14. Oelschlaeger, "Idea of Wilderness," 7.

15. Tom Birch, "The Incarceration of Wildness: Wilderness Areas as Prisons," *Environmental Ethics* 12 (1990): 7.

16. Barbara Allen, letter to the editor, *Northern Lights* 8 (1992): 29.

17. Evernden, *Social Creation*.

18. J. Baird Callicott, "La Nature est mort, vive la nature!" *Hastings Center Report* 22 (1992): 16-23.

19. Rolston, "Wilderness Idea Reaffirmed," 370 (emphasis added).

20. Ibid., 372.

21. Ibid., 377.

22. Callicott, "That Good Old-Time Wilderness Religion," 378.

23. Callicott, "The Wilderness Idea Revisited," 243.

24. Leopold, *Sand County Almanac*, 207.

25. N. Johnson et al., *Alternatives for Management of Late-Successional Forests of the Pacific Northwest*, report to the House Agriculture Committee of the U.S. House of Representatives, October 8, 1991.

26. J. Williams et al., "Declining Salmon and Steelhead Populations: New Endangered Species Concerns for the West," *Endangered Species Update* 9 (1992): 1.

27. U.S. Department of the Interior, Fish and Wildlife Service, "Status of Endangered Species Recovery Program Is Detailed in Report to Congress," *Endangered Species Technical Bulletin* 16 (1991): 9.

28. J. Ryan, "Conserving Biological Diversity," in *State of the World—1992*, ed. L. Brown et al. (New York: W. W. Norton, 1992), 13.

29. Michael Soulé, "The Millennium Ark: How Long a Voyage, How Many Staterooms, How Many Passengers?" *Zoo Biology* 5 (1986): 102-4.

30. Reed Noss, "Wilderness Recovery: Thinking Big in Restoration Ecology," *Environmental Professional* 13 (1991): 226.

31. Ibid., 243.

32. Callicott, "Wilderness Religion," 379.

33. Callicott, "Wilderness Idea Revisited," 239.

34. Craig Allin, *The Politics of Wilderness Preservation* (Westport, Conn.: Greenwood Press, 1982), 12.

35. Callahan, "Wise Use Movement," 14.

36. E. Brazil, "'Wise Use' Advocates Scorn the Rio Summit," *San Francisco Examiner*, 7 June 1992.

37. See F. Williams, "Sagebrush Rebellion," *High Country News* 24 (1992): 1, 10-11.

38. Brazil, "'Wise Use' Advocates."

39. Bryan Norton, *Toward Unity among Environmentalists* (New York: Oxford University Press, 1991), 58.

40. Gomez-Pompa and Kaus, "Taming the Wilderness Myth," 272.

41. Ibid.

42. Ray Dasmann, "Life-Style and Nature Conservation," *Oryx* 13 (1976): 281-86.

43. Gomez-Pompa and Kaus, "Taming the Wilderness Myth," 273.

44. Callicott, "Wilderness Idea Revisited," 242.

45. Rolston, "Wilderness Idea Reaffirmed," 375.

46. Paul Shepard and B. Sanders, *The Sacred Paw* (New York: Viking, 1985), 55-59.

47. Richard Nelson, *Make Prayers to the Raven* (Chicago: University of Chicago Press, 1983), 229-30.

48. Snyder, *Practice of the Wild*, 81-86.

49. Birch, "Incarceration of Wildness," 6.

50. For a full explanation of this platform see Ed Grumbine, *Ghost Bears: Exploring the Biodiversity Crisis* (Washington: Island Press, 1992), 184-228.

51. Salwasser, "Sustainability," 214.

52. Paul Shepard, "A Post-Historic Primitivism," in *The Wilderness Condition*, ed. M. Oelschlaeger (San Francisco: Sierra Club Books, 1992), 85.

53. M. G. Chandrakanth and Jeff Romm, "Sacred Forests, Secular Forest Policies, and Peoples' Action," *Natural Resources Journal* 31 (1991): 141-57.

54. Gomez-Pompa and Kaus, "Taming the Wilderness Myth," 273.

55. Michael Zimmerman, "The Blessing of Otherness," in *The Wilderness Condition*, ed. M. Oelschlaeger (San Francisco: Sierra Club Books, 1992), 269.

56. See Warwick Fox, *Toward a Transpersonal Ecology* (Boston: Shambhala, 1991), and Arne Naess, "Self-Realization: An Ecological Approach to Being in the World," *Trumpeter* 4 (1991): 35-42.

57. Snyder, *Practice of the Wild*, 37, 15.

2 . Habitable Earth: Wilderness, Empire, and Race in America

Denis Cosgrove

> The West of which I speak is but another name for the Wild; and what I have been preparing to say is, that in Wildness is the preservation of the World.
>
> *Henry David Thoreau*

The last phrase of Thoreau's famous proclamation has become the *locus classicus* of contemporary claims for the preservation of geographical areas and regions of wild nature. Rarely, if ever, do we consider the first part of Thoreau's sentence: "The West of which I speak is but another name for the Wild." Thoreau uses a geographical coordinate (a pure mapping concept with no essential environmental reference) synonymously with "the wild," and thus with wilderness. In so doing, he not only draws upon what I shall argue is a characteristically American mode of mapping the socioenvironmental geography of nationhood, but he also articulates in his own way key elements of a far broader discourse about chaos and order, identity and otherness, in which European culture has been engaged throughout the evolution of modernity. It is a discourse as much geographic and cartographic as it is environmental, as much social as it is natural, and as much about nation and empire as it is about individual and divinity. I shall argue that ideas of wilderness are deeply embedded in the creation of the modern world, but not exclusively so—myths of *garden* and *city* play an equally important role as landscapes of the imagination.

I shall treat modernity in the conventional historical sense as a European phenomenon, progressively apparent in all aspects of social, economic, political, and cultural life from at least the late fourteenth century and achieving global extension in the course of the twentieth century. As a geographer my specific focus is on two connected *spatial* dimensions of modernity. These are (1) the geographical expansion of the world known to and mapped by Europeans, and (2) the extension across that world of two characteristically European spatial forms of collective identity, dwelling, and authority: universal empire and nationhood. However we might theorize the engine of modern history, it has been the global projection of European power and European modes of life, together with the responses to these within and beyond Europe, that have produced the modern concepts of one world and a whole earth within which current wilderness debates are located.

The fifteenth-century Europeans who opened the era of overseas expansion and restitched the severed seams of Pangaea held an imaginative world map derived from classical and Islamic Mediterranean science, inflected with the beliefs of late Latin Neoplatonists and early Church fathers.[1] Represented by writers such as Macrobius, Isidore of Spain, and Albertus Magnus (and not challenged in its essentials by Ptolemaic cartography), the map was of a single landmass of three continents in the northern hemisphere of a spherical earth.

This was the *ecumene,* the habitable earth. Understanding of the Aristotelian *klimata* suggested that only the central band of this landmass, gathered around the "middle sea," should be truly habitable by civilized people. What existed beyond the ecumene—across the Ocean Sea, in the antipodes, in the boreal spaces of Scythia and Thule, or in the torrid zones of Libya— was unknown. Such spaces could not be the dwelling place of *men,* but rather of wild and savage nature, even of monstrous races, if Pliny and Solinus were to be believed. Those lands and their inhabitants lay beyond the bounds of civil dwelling, of *civilization,* whose etymological roots in Mediterranean discourse lie in the city with its landholding citizens *(cives),* the cultivators of their native territory.

Both classical and biblical traditions placed the city at the highest point in a hierarchy of imaginative environments built upon wilderness, that is, presocial, prediscursive space. A countervailing trope to the dominant theme of savage nature imagined that somewhere beyond the ecumene, in the East, might lie the terrestrial paradise, locus of a gentle, pre-lapsarian nature, humanity's innocent childhood. Under either interpretation, beyond the ecumene lay wilderness, which, whether savage or gentle, was a place where human dwelling, (i.e., garden and city) remained impossible because not subject to cultivation (i.e., culture).

The unknown, of course, always has to be constructed in the fashion of the known. Thus, the geographic discourse of wilderness beyond the ecumene was formulated dialectically out of Europe's own imaginative history and geography.[2] Two points are significant here. The first is the memory and experience of *empire.* In the European historical memory the ecumene had once been organized as a single *civis,* under Greece and later Rome. A continued claim to terrestrial and spiritual order was represented by emperor and pope within the idea of Christendom.[3] The second point is that the joint roots of late medieval European culture, classical and biblical authorities, concurred in narrating a progressive but cyclical history of socioenvironmental evolution from chaos to order, from wilderness through garden to city.[4] This discourse drew upon the Mediterranean experience of seasonal rhythms and the organic life cycle, mapping them jointly across time and space. United with the memory of empire, it gave rise to the belief that the history of civilization follows the course of the sun—toward the West.[5]

Renaissance aesthetic theory, drawing upon Aristotelian and Vitruvian sources, mapped social, corporeal, gender, and artistic orders into this spatiotemporal sequence. It gave Europeans a set of ideal types for the interpretation of their own landscapes and of those of the worlds they were encountering beyond the ecumene. It is not surprising that cosmographic and cosmological schemata illustrating this discourse were most fully elaborated and imaginatively represented during the sixteenth and seventeenth centuries, the peak of European geographic expansion.[6]

At this opening of the modern world, the European imagination had constructed a collective cultural identity out of historical narrative and geographic mapping. The historical narrative was of imperial order (Rome and Christendom) being imposed across the otherness of an intransigent wild nature. This could be *human* nature: original sin had to be cultivated out of the growing child. Environmentally speaking, it is found in the prehistory of natural wilderness, which had to be cultivated and civilized out of the barbarian world for civil dwelling to be possible. The geographic mapping was of civilized life within the city, supported by a surrounding garden of domesticated and cultivated nature, protected from the otherness of wild nature beyond.

Imaginatively, wilderness was at both the core and the periphery of this map, just as it was at both the beginning and the end of the historical narrative: it was original chaos, but it also remained the final act of the cycle of civilization, through the war and destruction that urban life and commerce inevitably brought. Wilderness was always correlated with origins and infancy, in the sense not only of innocence, but also of untamed human passions and undisciplined conduct. It also was correlated with Armageddon. The seeds of social development may be located in wilderness, but the wild itself is savage, animal, and presocial.

Anthropologists and sociologists today emphasize the necessity of the other in maintaining social identity among groups at all scales.[7] As their ecumene expanded to embrace an ever-greater part of the surface of the globe, Europeans were forced to adjust their imaginative histories and geographies to accommodate the empirical actualities they encountered at home and overseas. This process is well documented, especially for the most radically different of the geographic discoveries, the literal "New World" of the Americas.[8] I intend here only to identify certain aspects of it that relate to wilderness, and to do so for two defining moments: the initial discovery of America, and the final extension of imperial control over the territory of the most successful social and environmental cultural transplant of European Modernity, the United States.

Wilderness and the Caribbean Encounter

In the same months of 1492 during which Columbus was undertaking his historic voyage in the Ocean Sea, one of the most sophisticated projects of European publishing of the incunable period was under way in Nuremberg. Hartmann Schedel's *Chronicle,* published in Latin and German in 1493, reported the discoveries of Columbus, attributing them to Martin Behaim. A highly illustrated encyclopedic summary of late medieval knowledge, the work opens with a representation of the cosmogony from God's creation of order from the chaos of elements to a picture of the ecumene. The margins of this Ptolemaic map (see the illustration that begins this essay) are illuminated with a set of monstrous races, those wild others who existed beyond the margins of civilized dwelling. Half man, half animal, these creatures are most clearly marked as *in*human by the fact that they are *anthropophages.*

Today that word has been replaced as a name for eaters of human flesh by the word *cannibal.* The change is significant, for *cannibal* derives from the same root as *Carib,* and, as Peter Hulme has shown in *Colonial Encounters,* Europe's reading of the New World was from the beginning stretched between two discourses of wilderness tracing back to the Homeric epics. One of these posits the existence of the Arawaks, an innocent, childlike people inhabiting the islands of the Western Ocean during the Golden Age of harmony with the natural world. Their mortal enemies were the Caribs, scarcely human wild warriors and cannibals bent on conquering, enslaving, and (where appropriate) eating the gentle Arawaks. Such is the enduring power of this narrative that it still informs the historical anthropology of the Caribbean. The historical accuracy of this ethnography is disputed, but hardly significant for our purposes because, as Hulme notes:

These are ideological, not historical categories and as such have a long history, but the very shock of the contact between Europe and America gave the couplet a new lease of life. These days there are few human societies unaffected by the expansion of Europe and its aftermath, yet ideologically the couplet's demise seems capable of almost infinite postponement. There is always at least the rumour of a last 'primitive' society, inevitably cannibalistic, unvisited by camera or notebook and which, when visited, turns out to have renounced cannibalism only recently. Fortunately, just beyond the next hill is another society, unvisited as yet by anthropologists, and they are *still* cannibals. Exactly the story Columbus heard nearly 500 years ago.[9]

Cannibalism is, of course, the ultimate act of the wilderness condition, the very antithesis of civil society, as the fate of the Franklin and Durren expeditions would later remind "civilized" Europeans. However, more often than not, a different angle of the modern gaze uncovers gentle peoples of the rainforest, imaginative descendants of the Arawaks, representatives of humanity's deep past, living in supposed Paleolithic harmony with their wilderness environment.

The application of this Odyssean Mediterranean discourse to the Atlantic discoveries extended during the sixteenth and seventeenth centuries to the whole New World. The course of empire took its way westward, and at every stage the wilderness and its occupants acted as the counterpoint to a constantly reshaping European identity. Attempts to create and hold an imagined middle landscape, neither savage wilderness nor bloated city, run like a yearning refrain—"back to the garden"—through the centuries of American conquest and incorporation into the European *imperium*, all the way from its settlement by Europeans and its coming to self-consciousness as a nation, through Jefferson's Land Ordinance Act, and on to the contemporary suburb.[10] But it is a refrain constantly drowned out by drums of war proclaiming the need to conquer the wilderness and its savage inhabitants, while sweet pastoral pipes hymn elegiac strains of lost innocence and purity in a once-perfect nature and among its now-destroyed aboriginal peoples.

Nationhood and Wilderness

As Europe's overseas imperium expanded, so the territorial geography of the old continent was internally reshaped into competing nation–states. Wilderness, garden, and city are repeatedly mapped into the imaginative geography of European nationalism. Romantic nationalism is apparent in John Ruskin's celebrations of harmony between peasant cultures and the natural world.[11] It is a driving force behind the topographic mapping and recreational exploration of European naturescapes from the early nineteenth century until well into the twentieth century. Oswald Spengler's reference to cultures that grow with original vigor out of the lap of a maternal natural landscape, to which each is bound during the course of its existence derives from the intense German concern to map national time and space.

In Britain, the national (and imperial) capital of London is mythically surrounded by the "garden of England"—its Home Counties where the true social character of the nation is said to find expression in a domesticated landscape. Beyond, at the wild and upland margins of the kingdom, are the haunts of ancient Britons, half-savage Celts, Scottish, and Irish in whose loins lie the seeds of British nationhood, whose physical prowess made them suitable to form the fiercest regiments or the advance pioneers of empire.[12] Today in these "marginal" lands are to be found the national parks, areas of upland wilderness to be preserved for the nation.[13]

The national park is of course an American invention, subsequently adopted by other nations. Its historical and intellectual contexts have been more than amply documented.[14] In the United States, and from very early in other parts of the extended European ecumene, national parks have occupied areas regarded as wilderness, the last preserves of places untouched by the outward expansion of European imperium.[15] In most such areas former modes of dwelling by indigenous peoples had been expunged, often violently, not many years before the declaration of wilderness status, or would be when that designation was made.[16] In the same years the radical imposition of new modes of dwelling found its highest expression in all these areas of European settlement in the construction of nations, a form of territorial social organization that had emerged in Europe from the internal ruins of empire.

The late nineteenth century, between the unifications of Germany and Italy in 1870 and the outbreak of continental war in 1914, saw the flood tide of European nationalism and imperialism as well as a growing national self-awareness among temperate colonial societies.[17] One cultural dimension of this phenomenon was the interest in the origins of the nation, an interest that had its roots in late-eighteenth-century Ossianism and romanticism and that was critical for providing a mythical pedigree for national difference. Archaeology, folk revivals, and the recovery—often reinvention—of customs, traditions, and costumes were the apparently innocent dimension of a nationalist obsession whose darker side was revealed in the scramble for empire in Africa, in European eyes the last childlike/savage wilderness to be brought into the light of civilization.

Among the many ideological pressures that stamped the different aspects of European nationalism into a single coin was the belief in race and eugenics. The success of Darwinian biology, particularly after Mendelian genetics furnished a clue to the mechanism for the inheritance of acquired traits, gave a huge stimulus to racial theories of cultural difference that provided an apparently scientific foundation for the differentiation of European peoples into distinct nations. The issue of internal social origins in the prehistoric European wilderness became as significant as differentiating the world's races into the evolutionary hierarchy of civilization in justifying the European claim to global authority through empire.[18] At the turn of the century a complex set of discourses connected human and natural origins through Darwinian ideas of human evolution linked with the greater primates by Huxley, Polar and African exploration, and gender stereotyping.[19]

The United States, since 1776 the very model of a European Enlightenment state, shared these concerns for defining national origins and identity.[20] In the aftermath of the Civil War, the United States completed the conquest of its own indigenous people and became a modern industrial nation equivalent to Great Britain, France, or the emerging Germany. Grounding national identity in a nation of immigrants from across Europe, many still first-generation Americans, was more complex in the United States than in Europe. The question of origins could not be settled internally through folklore, archaeology, and Dark Age historical myths, for the truly aboriginal peoples were seen as nondwellers because they were believed to be noncultivators. Regarded as inferior racial types, they were physically removed into an apartheid system of reservations.

In the resolution of American national identity, wilderness was to play a significant role, though rather different than it did in Europe. In Europe wilderness existed in time; in America it could be found in space. Frederick Jackson Turner's 1893 paper, "The Significance of the Frontier in American History," gave western wilderness the pivotal role in the building of American national identity. The thesis states that in the confrontation between civilization and savagery on a westward-moving line of settlement, European social sophistication was to be stripped away and replaced by a healthy young democracy. The final words of Turner's essay, "the frontier has gone, and with its going has closed the first period of American history," reflect the fear that the second century of American nationhood could no longer enjoy the benefits of this sociogeographic process.[21]

It was a fear less of the frontier, either wild or newly cultivated, than of events concurrently taking place in the American city. Turner's thesis encouraged a surge of interest in the preservation of western wilderness that reached its peak in the designation of national parks during the presidency of Teddy Roosevelt.[22] The further link to debates about American racial origins has, I believe, been less noticed.[23] Yet it is this link that allows us to place the American mapping of wilderness more centrally into the long history of European geographic imaginings and to challenge claims to American exceptionalism.

Wilderness and Identity in Early-Twentieth-Century America

More significant than the closing of the frontier in the minds of most Anglo-Americans in the 1890s was the issue of immigration. Between 1890 and 1914 some fifteen million new immigrants entered the United States, cresting at 1.29 million in 1907, a figure not exceeded until the 1980s.[24] In that year Congress set up the Dillingham Commission to investigate the situation. The commission's report was based upon a single key belief: that since 1880 a new type of person had come to dominate movement into the United States. Unlike the "old" immigrants from northwestern Europe, who for the most part had arrived in family groups, entered farming (thus truly dwelling on the land), and "mingled freely with . . . native Americans [sic]," the new immigrants were unskilled, transient young men, largely from southern and eastern Europe, entering urban industrial employment and keeping a distance from earlier settled Americans.

In making such a sharp contrast between the two groups the report was incorrect in every respect but two. The great wave of immigrants after 1880 did enter urban industry (but less by choice than because that had become the growth sector of the American economy). Also, the vast majority (81 percent) did indeed come from new European origins: Austria–Hungary, Romania, Russia, Italy, Greece, and Turkey, often removed from their native lands by the ethnic cleansing under way there.

The Dillingham Commission was largely a response to the pressures of an American nativism born out of Anglo-Saxon fears of "race suicide" in the face of urban disease and recent immigrants' reproductive capacity.[25] This nativism grew rapidly from the 1880s, most virulently in the Pacific West, where it was directed against Chinese, Japanese, and other non-Europeans. In more genteel New England it took the form of Immigration Leagues led by Boston intellectuals such as Harvard geologist Nathaniel Shaler, himself an academic conqueror of western wilderness.[26] In their writings, nativists sought to root American national origins in an Anglo–Saxon ethnicity under which they grouped British, Germans, Scandinavians, and (reluctantly) Irish, while excluding Italians, Jews, and Slavs as lacking the "ancestral experience" that characterized established Americans. By 1906 the Immigration Leagues were using explicit geneticist arguments to answer the question of whether America should "be peopled by British, German and Scandinavian stock, historically free, energetic, progressive, or by slow Latin and Asian races, historically downtrodden, atavistic, and stagnant."[27] Their arguments increasingly revolved around a belief in racial purity as the foundation of American national greatness, although some did suggest, following the Turnerian line, that "racial and hereditary habits" might be overcome by exposure to the American environment.[28]

In such a context, the great surge of wilderness preservation in the first decade of the new century (so often attributed to the individual charisma of men such as John Muir and Teddy Roosevelt) should perhaps be reassessed. The open spaces of the West were as far removed geographically, culturally, and experientially from the crowded immigrant cities of New York, Philadelphia, Pittsburgh, or Chicago as they could be. Patrons of the wild came overwhelmingly from "old stock" and the middle classes; many had been as strongly committed to the introduction of urban parks modeled upon the English picturesque tradition (a style often opposed by newer immigrants, who favored playgrounds and baseball diamonds to lakes, trees, and flower beds). Furthermore, the national parks represented, in Turnerian terms, the kind of environment in which earlier—and racially purer—immigrants were believed to have forged American national identity. Western wilderness was not merely the theater of American empire, it was the idyll of America's national childhood, wild, innocent, and free. In its preservation lay the preservation of the nation's youthful hope.

Youth and innocence are, as I have demonstrated, long-standing features of wilderness ideology. They have been important themes in American cultural discourse from the moment of European discovery, but they achieved a special prominence in these same *fin-de-siècle* years as new immigration and national park legislation. Nineteenth-century romanticism had encouraged a positive reading of childhood as a separate sphere of human experience. Childhood was considered close to chaos, thus requiring the civilizing process. Simultaneously, like femininity in late Victorian bourgeois society, it was untainted by social artifice and so embodied a "moral innocence and emotional spontaneity which seemed increasingly absent from the public realm."[29] In the later years of the century the cult of childhood merged on the one hand with the cult of strenuous exercise as a route to the recovery of unrepressed emotions and the imaginative life, and on the other with ideas of racial purity. As an 1896 reviewer in *The Nation* put it:

> Scientists have informed us that the children alone possess in their fullness the distinctive features of humanity, that the highest human types as represented in men of genius present a striking approximation to the child type.[30]

In order to sustain the quality of the race in adulthood, eugenically informed educationalists recommended repetition during childhood of the experiences and emotions of primitive ancestors and the cultivation of primitive vitality. Where better to achieve this than in the wilderness of unspoiled nature, especially for Americans whose ancestral heroes were supposedly pioneers and frontiersmen?

At the century's turn it was also widely believed that physical prowess and willpower were intimately connected. The national heroes of both European imperial nations and the United States were manly pioneers and explorers engaged in conquering the final outposts of empire, big game hunters directly confronting enraged male primates in Africa or India, or racing to be first to plant their flags at the ends of the earth.[31] These were the role models for youthful members of patriotic and character-forming groups such as the Scouts and Woodcraft Folk. The president himself incorporated and articulated these values; Teddy Roosevelt was a devotee of African safaris and his visits to the Yosemite wilderness with John Muir were not so dissimilar. It was Roosevelt who warned fellow Americans that a race was "worthless and contemptible if its men cease to be willing and able to work hard and, when needed, fight hard, and its women cease to breed freely."[32]

Western national parks may not have been set up to be breeding grounds for a particular type of American national character, but the cultural context within which wilderness gained respect in America is important in understanding its implications. That context not only suggests a closer link with imperialist, xenophobic, and racist features of American nationalism than many Americans would feel comfortable espousing today, it also locates the discourse of wild nature in the early national park movement within a historically deep-rooted cultural tradition that embraced ideas of order, civilization, empire, and both personal and national self-consciousness. Contemporary appeals to the idea of wilderness still retain these hidden attachments.

Wilderness Ideas Today

Today, designated wilderness preserves in the United States occupy some 80 million acres, or 3.5 percent of the national territory, largely the result of sustained pressure upon the federal government since the 1930s by environmentalist groups such as the Sierra Club, the National Audubon Society, and the Wilderness Society. Membership of these groups has been closely tied in personal and intellectual pedigree to nineteenth-century nature advocates pressing for urban and later national parks to improve the material and spiritual health of urban America.[33] A sustained but empirically unsubstantiated belief in the spiritual power of wilderness as an antidote to the terrible divorcement from a native environment supposedly occasioned by urban living has remained central to these groups arguments for wilderness at least until the past decade, and continues even today in some quarters.

To the casual observer, the most striking sociological feature of any visit to an American national park or forest, and even more to a designated wilderness area, is how little of the vibrant social and ethnic mix of America is represented by the visitors. Young, white, above all healthy, if not wealthy, middle-class families overwhelmingly dominate the camping sites and visitor centers,[34] while the hikers and backpackers on the wilderness trails usually are youthful members of the same class and ethnic group that the Immigration Leagues of the early part of this century had proclaimed original Americans. It is hardly surprising that such people should be young, fit, and well-off: the arduous physical exercise necessary is unlikely to appeal to the elderly and infirm, while the costs of access excludes the poor. But the highly elaborated codes of conduct and dress for these areas can be as rigid and exclusive in their moral message as in their expense. They articulate an individualistic, muscular, and active vision of bodily health largely derived from Anglo sociopsychological culture. The literature of wilderness experience suggests that many features of the American cult of nature have remained little changed since the early twentieth century.

At the more scholarly level, wilderness writing no longer stresses the direct effects of nature experience on individual and social health, although its beneficial impact upon personal psychology is still proclaimed. The cutting edge of the argument for wilderness has shifted toward a progressive rhetoric borrowed from civil rights and liberation movements and extended beyond humans to nature. This extension is predicated on the belief that evolutionary theory has undermined any philosophical or theological arguments for human exceptionalism within the created order. Wilderness thinking remains profoundly colored by a teleology of natural order, and indeed by a desire to read the social into a narrative of the natural.

The argument also remains dominated by an uncritical acceptance of a bio-ecological paradigm whose intellectual roots are tangled up with much of the unsavory racial and eugenic theorizing of the early twentieth century.[35]

Max Oelschlaeger, for example, in a work that in every other respect is profoundly skeptical of received wisdom and that self-consciously proclaims the postmodern collapse of scientific foundationalism, treats evolutionary theory as foundational and the ecological discourse of ecosystems and entropy as a mimetic representation of nature rather than as a currently accepted metaphor whose half-life is unlikely to be longer than that of the mechanical metaphor he so properly condemns.[36] His own historical narrative, in its movement from the "deep time" of the Paleolithic to the forecast future of global ecocatastrophe via the Neolithic garden and modernist city, follows the well-worn trajectory of the West's mythical narrative of social evolution that we have examined here.

Perhaps the most disturbing aspect of much contemporary wilderness ideology is its continued attachment to the mentality of imperium. I have sought to emphasize the specifically American construction of the modern wilderness idea while placing it in the long history of European imaginative mapping and dwelling in the world. Since the early 1970s one of the most popular icons of the wilderness movement has been a photograph of the earth taken during the 1972 Apollo 17 lunar mission. In Green movement literature especially, this image has replaced conventional cartographic representations of the earth based upon the graticule of coordinating latitude and longitude—from the late fifteenth century the tool and image of European expansion and spatial control. Apollo's elemental picture of earth, water and atmosphere, the very image of Gaia, originated as a celebration of American imperial reach, produced en route to the act of placing the U.S. flag on the moon, in the time-honored motif of imperial claim-staking.[37]

Subsequent interpretation of this ecumenical image by environmentalists as revealing a vulnerable earth, dominated by nature and isolated in the secular void of dead space, draws upon specifically American yearnings for socioenvironmental perfection, isolation, and the need for moral regeneration through recognition of nature's claims. Above all, this interpretation of "whole earth" serves increasingly to justify what are effectively imperialist interventions anywhere across the globe in favor of environmental and wilderness preservation. At the very historical moment when destruction of the old European imperium has finally extended the system of territorial states and national integrity pioneered by European (though also somewhat imperialist) Enlightenment beliefs in human rights, social perfectibility, and the creation of the liberal state across the globe, it is in danger of being undermined by the ideology of an empire of nature. The maintenance of environmental order in this empire is policed by spokespeople and activists drawn from the old imperial nations, often using the same language of pioneering adventure—and in the same theaters of snow, desert, and jungle—as their colonial forebears.

Conclusion

The wild and the West were intertwined within the European imagination for centuries before Thoreau, but less as a place of human dwelling than for mapping visions of social and environmental order. The setting of wilderness remains a representation of contemporary social and psychological formations, not a bedrock reality. In a global political economy, with its consumerist creation of meaning, the commodification of wilderness parallels that of the city. The fantasies traditionally associated with wilderness today produce phenomena such as an urbanized, traffic-clogged Yosemite, international "cannibal tours," and Everest's waste disposal problem.[38] Finally, historical understanding should encourage caution—the strongest voices for wilderness come once again from the West, from an America forced to question its national identity in the face of mass immigration. The ideology of wilderness is a potent weapon in social discourse. It needs to be treated with great care.

Notes

1. B. Harley and D. Woodward, *The History of Cartography,* Vol. 1 (Chicago: University of Chicago Press, 1987).

2. Oelschlaeger prefers to see this complex of ideas as an atavistic social memory from the Neolithic, and contrasts it to a Paleolithic "idea of wilderness." Such an interpretation is understandable in view of the biological interpretation of humanity upon which his thesis rests. There is no empirical warranty for this Paleolithic mind, and in my opinion the thesis radically underestimates the creative power of human imagination working with historically changing material circumstances. See Max Oelschlaeger, *The Idea of Wilderness* (New Haven: Yale University Press, 1991).

3. S. J. Edgerton, "From Mappa Mundi to Mental Matrix to Christian Empire," in *Art and Cartography: Six Essays,* ed. D. Woodward (Chicago: University of Chicago Press, 1987).

4. See N. Frye, *The Great Code* (Princeton, N.J.: Princeton University Press, 1978). The cycle of civilization, progressing through "ages" of gold, silver, bronze, and iron, forms the common element of Hesiod's *Works and Days;* Ovid's *Metamorphoses,* and Virgil's triad of *Eclogues,*

Georgics, and *Aeneid*. See, among others, Clarence Glacken, *Traces on the Rhodian Shore* (Berkeley: University of California Press, 1967).

5. See L. Barritz, "The Idea of the West," *American Historical Review* 66 (1961): 618-40, and J. H. Elliott, *The Old World and the New 1492–1650* (Cambridge: Cambridge University Press, 1992), 94.

6. Denis Cosgrove, *The Palladian Landscape* (Leicester: Leicester University Press, 1993).

7. B. Anderson, *Imagined Communities* (London: Verso, 1983).

8. A. Pagden, *European Encounters with the New World: From Renaissance to Romanticism* (New Haven: Yale University Press, 1993); Elliot, *Old World and New*; E. Zerubavel, *Terra Cognita: The Mental Discovery of America* (New Brunswick, N.J.: Rutgers University Press, 1992).

9. P. Hulme, *Colonial Encounters: European and the Native Caribbean 1492–1797* (London: Routledge, 1986), 83.

10. D. Cosgrove, *Social Formation and Symbolic Landscape* (London: Croom Helm, 1984); P. Rowe, *Making a Middle Landscape* (Cambridge: MIT Press, 1991).

11. D. Cosgrove, "John Ruskin and the Geographical Imagination," *Geographical Review* 69 (1979): 43-62.

12. It is significant in this context that in Theodore de Bry's *America*, his illustrations for John White's account of the English settlement of Virginia should make an explicit and graphic connection between the aboriginal peoples of the New World and Ancient Britons. See also D. Cosgrove, "The Picturesque City: Nature, Nations and the Urban Since the Eighteenth Century," in *City and Nature: Changing Relations in Space and Time*, ed. T. Kristensen et al. (Odense: Odense University Press, 1993), 44-58; P. Taylor, "The English and Their Englishness: 'A Curiously Mysterious, Elusive and Little Understood People,'" *Scottish Geographical Magazine* 107 (1991): 146-61.

13. A comparative research project funded by the European Community currently is examining the connections between European nationalism and the culture of nature in Italy, Denmark, Sweden, and Great Britain during the first two-thirds of the twentieth century, a period that includes both the most virulent expressions of nationalism in Europe and the establishment of national parks in most Western European national territories.

14. R. Nash, *Wilderness and the American Mind* (New Haven: Yale University Press, 1983); D. Worster, *Nature's Economy* (Cambridge: Cambridge University Press, 1977); A. Wilson, *The Culture of Nature: North American Landscapes from Disney to the Exxon Valdez* (Cambridge, Mass.: Blackwell, 1992).

15. See Wilson, *Culture of Nature*, and J. MacKenzie, *The Empire of Nature* (Manchester: Manchester University Press, 1987).

16. See R. F. Neumann, "Field of Dreams: Colonial Recasting of African Society and Landscape in the Serengeti National Park," *Ecumene*, forthcoming.

17. E. J. Hobsbawm, *Nations and Nationalism since 1780: Programme, Myth, Reality* (Cambridge: Cambridge University Press, 1990).

18. D. N. Livingstone, *The Geographical Tradition: Episodes in the History of a Contested Enterprise* (Oxford: Blackwell, 1992).

19. Donna Haraway, *Primate Visions: Gender, Race and Nature in the World of Modern Science* (London: Verso, 1992).

20. The United States was the first nation to define itself primarily in terms of the land, and only secondarily in terms of the people. This was inevitable in a territorial state whose existence derived from land settlement by peoples of diverse languages, religion, and ethnicity. The point was observed in the earliest years of the republic by St. Jean de Crèvecoeur in his famous essay, "What Is an American?" See also A. Kolodny, *The Lay of the Land* (Chapel Hill: University of North Carolina Press, 1975).

21. G. Kearns, "Closed Space and Political Practice: Frederick Jackson Turner and Halford Mackinder," *Environment and Planning, Society and Space* 2 (1984): 23-34.

22. Nash, *Wilderness and the American Mind;* B. Novak, *Nature and Culture: American Landscape and Painting 1825–1875* (New York: Oxford University Press, 1980); S. Daniels, *Fields of Vision: Landscape and National Identity in England and the United States* (Princeton, N.J.: Princeton University Press, 1993).

23. R. Horsman, *Race and Manifest Destiny: The Origins of American Anglo-Saxonism* (Cambridge, Mass.: Harvard University Press, 1981); R. J. Park, "Biological Thought, Athletics and the Formation of a 'Man of Character,'" in *Manliness and Morality in Britain and America 1800–1940,* ed. J. A. Mangan and L. Walvin (Manchester: Manchester University Press, 1987), 1-28; Haraway, *Primate Visions.*

24. M. A. Jones, *American Immigration* (Chicago: University of Chicago Press, 1960).

25. Park, *Biological Thought.*

26. D. N. Livingstone, *Nathaniel Southgate Shaler and the Culture of American Science,* (Tuscaloosa: University of Alabama Press, 1987).

27. Prescott F. Hall (1894), quoted in Jones, *American Immigration,* 259.

28. M. Bassin, "Turner, Solov'ev, and the 'Frontier Hypothesis': The Nationalist Significance of Open Spaces," *Journal of Modern History* 65 (1993): 473-511.

29. T. J. Jackson Lears, *No Place of Grace: Antimodernism and the Transformation of American Culture 1880–1920* (New York: Pantheon, 1981), 146.

30. Ibid., 147.

31. J. MacKenzie, *Propaganda and Empire* (Manchester: Manchester University Press, 1984); Y. F. Tuan, "Desert and Ice: Ambivalent Aesthetics," in *Landscape, Natural Beauty and the Arts,* ed. S. Kemal, and I. Gaskell (Cambridge: Cambridge University Press, 1993), 139-57; B. Lopez, *Arctic Dreams: Imagination and Desire in a Northern Landscape* (New York: Bantam Books, 1987).

32. Quoted in Park, *Biological Thought.*

33. T. Young, "*Viewing the Wilderness, or Recognizing Environmental Determinism in Two Environmental Movements,*" paper delivered at the annual convention of the Association of American Geographers, Atlanta, April 1993.

34. The median age of national parks and wilderness area visitors is slowly rising and currently is slightly above forty years. The reasons for this are unclear, but relate in part to the baby-boom generation passing through the American population. This bulge group came of age in the years surrounding passage of the Wilderness Act and participated in the first wave of ecological consciousness during the late 1960s.

35. Ernst Haeckel, social Darwinist and one of the fathers of ecological science, was also one of the earliest to argue that nature had rights. Some more recent supporters of bioregionalism and deep ecology sometimes betray attitudes toward society similar to those of early-twentieth-century eugenicists, especially in their views on human population control.

36. Oelschlaeger, *The Idea of Wilderness.*

37. See D. Cosgrove, "Contested Global Visions: One-World, Whole-Earth, and the Apollo Space Photographs," *Annals, Association of American Geographers* 93 (1994).

38. D. McCannell, "Nature Incorporated" and "Cannibal Tours," in *Empty Meeting Grounds: The Tourist Papers* (London: Routledge, 1992).

3 . Earth-Talk: Conservation and the Ecology of Language

Max Oelschlaeger

When I recently read E. O. Wilson's *The Diversity of Life*, it was with a grow-ing sense of approval, since I came to realize that here was precisely the kind of scientist that Arne Naess believes essential to the future, a man who dares to say what he knows is true and feels is right, a man who provides "articu-late leadership," to use Naess's words.[1]

We verge, Wilson argues, on a *mass extinction of life caused by nothing more and nothing less than human action.* His book gives "evidence that humanity has initiated the sixth great extinction spasm, rushing to eternity a large frac-tion of our fellow species in a single generation." He argues relentlessly that "every scrap of biological diversity is priceless, to be learned and cherished, and never to be surrendered without a struggle."[2] Wilson estimates that the loss of species has increased from a minimum of 1,000 per year during the 1970s to more than 10,000 per year.[3] I emphasize that this is a conservative es-timate, the best possible face that can be put on ecocatastrophe.[4]

Those are the facts.[5] The question I address in this essay is more than any-thing one of strategy for the conservation of wild places and the indigenous flora and fauna. How is it, I ask, that species *Homo sapiens* seems incapable of acting—even in its own self-interest—to preserve the diversity of life on earth? What impels Western culture to continue cutting the figures of an ecological *danse macabre?* Why is it that the wondrous and mysterious chaos of life, the wildness that lies within our own bodies and that animates the living creatures covering earth, is subject to the relentless forces of domesti-cation? What can be done to constrain the growing tide of humanity that sweeps across the earth, enveloping and inevitably destroying the precious remnants of wilderness in its wake? And how is it that the idea of wilderness can be translated into a meaningful framework for the conservation of life?[6] Give me a wilderness that no civilization can withstand, wrote that wild iconoclast, Henry David Thoreau. In wildness is the preservation of the world.

Consider, even hypothetically, the "bookend" premises that communication is culture and that culture is communication. By doing so we gain perspective: a path appears before us, the fateful route that our kind treads, the road of economic development that extends itself across the planet, crisscrossing even remote areas of Alaska and Brazil. Of course, the impulse to develop has a new name these days, a linguistic veneer that conceals the reality that the human species is, through relentless economic development and population growth, precipitating a mass extinction of life.

That name is *sustainable development,* now a buzzword among mainstream (reform) environmentalists. One knows that the term is far from the cutting edge when the president of the United States creates a Council on Sustainability, replete with many appointed members who are CEOs of Fortune 500 companies. Little matter that economic development by any name is better termed "maldevelopment," as Vandana Shiva points out with admirable honesty.[7] We will manage earth and secure "our common future"— that is, extend the process of high-tech, industrialized economic growth to every person in every land so that all 5.6 billion human beings can be just like Europeans and Americans and Japanese, caught in a glut of materialism and hyperconsumption. Through politics and technology our common future is apparently a sure thing, since those who lead the international conservation movement can find "no limits to growth."[8]

Culture, I repeat, is communication. But who controls global communication? According to those who study these things, we live in the midst of an information revolution based on technologies controlled by multinational corporations whose primary interest is in globalizing themselves. Could it be that communication is part of the technology of power that promotes the status quo,[9] protects the prevailing order of things, and co-opts the conservation of wild lands and creatures in the name of sustainable development?

Consider Paul Kennedy's recent, remarkable book, *Preparing for the Twenty-first Century.* His voice is much like that of a lone wolf howling in the wilderness, but his words are worth repeating. He finds the "vision of a prosperous and harmonious world economic order" nothing less than "breathtakingly naive in light of this planet's demographic, environmental, and regional problems." Indeed, he continues, there is a failure among our governing elites (political and economic, both of whom are well represented here) "to recognize that newer technologies may not benefit all, that the vast majority of the world's population may not be able to purchase [consumer] goods . . . , and that profound changes both in economic production and in communications can bring disadvantages as well as advantages in their wake."[10]

Perhaps, just maybe, there is another way to think about conservation, a way outside of the prevailing vocabulary of reform environmentalism. Let us redirect our gaze along the road less traveled, away from the talk of sustainable development and our common future, and toward the past. An event, actually more a sequence of events, looms fatefully along the rearward trajectory of culture. Literacy. The shift from orality to literacy is a fateful event in the history of the human species. We are embedded in the communication media of a literate, advanced industrial culture. The alphabet and all that trails in its wake, including the mechanically reproduced, printed word, I am saying, is a technique that has untold social, psychological, and ecological consequences.[11] Above all, literacy has separated us from earth, from any sense of kinship with the larger community of life from which we came and in which we remain embedded. And it has helped to sustain the process of domestication (initiated by agri-culture) that now threatens to consume the planet.[12]

How, you might ask, is my line of inquiry relevant to the task of protecting biodiversity, or, more generally, translating wilderness into a lingua franca for the conservation of biodiversity in the twenty-first century?[13] By all indications it is paradoxical to concentrate on language, that most human and civil of all things, since it apparently obliterates the wilderness. By common sense alone it appears that we can protect the wilderness if and only if we use language and the media, as communication.

Common sense, it has been remarked by many, is the most uncommon sense of all. There is no paradox in focusing on language, for therein lies the route back to *the source,* to the wildness in which lies the preservation of the world. Marjorie Grene argues that "we still have the image of a human world shorn of any roots in nature and a natural world devoid of places for humanity to show itself." Accordingly, we need to show from within language itself, that is, in a fully reflexive way, how "historicity, as necessary condition for, and defining principle of, human being, can be within, not over against, nature."[14] In other words, if it is through language that we have been alienated from nature, then reconciliation might also be effected through language—through *earth-talk.*

As a variety of scholars have made clear, though the human animal has been a language animal almost from the beginning, literacy (followed in turn by typography and the electronic revolution in communication) changed the way humans were in the world, the way in which they conceived of themselves, and the way in which culture itself was sustained.[15] Standing on the far side of literacy, it is difficult for us to imagine, let alone grasp, the reality of life in a primary oral culture. Calvin Martin's recent *In the Spirit of the Earth*, a brilliant attempt to rethink the categories of history and time, offers a useful starting point. Primary oral peoples, Martin contends, kept alive in their discourse a sense of "intimate identification, indeed kinship, in a mythically literal sense, with the rest of earth's society—a speech and technology of power, though not so much a power *over* other things as the power *of* these other sentient beings."[16]

So conceptualized, the oral peoples of the earth did not face the problem that ecological-minded scholars across a wide array of interface disciplines (eco-history, eco-economics, eco-ethics, and so on) believe must be addressed if we are to conserve biodiversity, namely learning how to heal the fissure between culture and nature, between our sense of history and its processes and our sense of nature and its processes. As Carolyn Merchant puts the point, "Only by according ecology a place in the narrative of history can nature and culture be seen as truly interactive. . . . [American] Indians [and other primary oral peoples] exemplify the true meaning of ecology as *oikos* (house) by making the entire habitat . . . their home."[17] Earth-talk is the way in which we might find our way back to what we have lost, not in the sense that the mass of humanity will return to hunting and gathering, for that is an impossibility, but in the sense that we will overcome the alienation, inherent in literacy, from the community of life. As Paul Shepard contends, if we recover Paleolithic sensibilities, now concealed behind the theorizing of difference (a characteristic gesture of literate people), then those threads can be woven into our own culture, since culture is a mosaic—that is, never finished, once and for all.[18]

Let us proceed along the path slowly. The hold of literacy on the modern mind is stronger than it appears. Indeed, we are language animals, storytelling culture-dwellers whose specifically human beingness is enframed by language. Neil Evernden argues that the common core of the dominant ideology concerning the relationship between wilderness and civilization is *environmentalism itself*. "We call people environmentalists because what they are finally moved to defend is what we call environment. . . . Ironically, the very entity they defend—environment—is itself an offspring of the nihilistic behemoth they challenge. It is a manifestation of the way we view [or theorize, from inside the frame of literacy] the world."[19]

Environmentalism, in whatever guise, is a way that literates theorize the world, a worldview that totally enframes our perception, valuation, and categorization of wilderness and its relation to us as human beings. Environmentalism as ideology is the theorizing of difference in a fashion that draws an uncontested boundary between the human and the so-called nonhuman.

More specifically, environmentalism perpetuates the legitimating narrative that separates culture (including history, society, and personality) from nature (including land, air, water, and other creatures). Ultimately, nature, now conceptualized as environment, appears to be little more than the stage upon which the main show, that is, the human drama, is performed.

The fence between nature and culture is implicit in the ideology of sustainable development, which aims above all else to extend the process of domestication so that all the earth might be harnessed to exclusively human purpose. Concealed behind the facade of sustainable development are two facts:

First, the facade conceals the fact that managing earth is, on the balance, a mission impossible. John Firor, former director of special projects at the National Center for Atmospheric Research, argues that at best there are "many uncertainties" inherent in the project to manage the planet, since it

> requires not only that we replace the evolved systems on which we depend, but also that we be wise enough to do so in the right sequence and completely, so that at no time during the process we are left without food to eat, air to breathe, and strong governments to keep the peace. This qualification is needed because the increasing domination of natural systems by our expanding technological society also means that in addition to freeing ourselves from our dependence on these systems, we are also causing their gradual disappearance. We will need to replace the accumulated "wisdom" of the interconnections between air, earth, water, and species with our own intelligence, diligence, and management skills.[20]

A second fact concealed by the ideology of environmentalism, according to E. O. Wilson, is that "by every conceivable measure, humanity is ecologically abnormal."[21] Sustainable development is more than anything else a psychological subterfuge by which our political and economic elites avert their vision from the sordid reality of the relentless growth of human population and the spread of industrialism grinding life beneath its heel across the planet.

How might we begin to change the course of events that initiated and sustains the extinction of biodiversity? Evernden invites us to abandon that "nihilistic behemoth," the modern worldview that underlies international reform environmentalism. And yet, paradox of paradoxes, we are people who presumably must think of the world in terms of the learned categorical scheme of environmentalism. There appears to be no possibility of an alternative language, for to create one appears to entail abandoning the cultural project upon which we have been so long embarked. From this fundamental human predicament there is seemingly no escape. And yet there is the possibility of understanding something beyond environmentalism for those who are willing to open themselves to the ecology of language, to earth-talk.

Where are the words, Thoreau asks, that speak for nature, that still have earth clinging to their roots? Speak we must, as I am doing now. Yet in our characteristic speaking we have silenced wild nature, the biophysical chaos. We have become MAN. MAN stands over nature, categorizing nature, converting nature, harnessing the wilderness to the productive project of civilization. His voice rules.[22] MAN believes that he is in control of the planet, that through technological fixes, environmental engineering, and biotechnology he can continue upon the trajectory of modern history.

Each of us finds our place in the project that is MAN—butcher, baker, candlestick maker . . . philosopher, environmentalist, economist, and politician, too. We are all MAN, caught up in the historic project of the West. (My intent here is not to efface my Third World brothers and sisters. I feel your gaze, your presence, and your pain. We in the West, in the First World, have more to learn from primary oral cultures about living on earth than you have to learn from us. But that is another story.)[23]

Of course, there are among us poets, writers, and composers who give voice not to MAN but to brother earth and sister moon. I think of Gary Snyder, whose poetry creates a dwelling space, an old–new eco-social terrain named Turtle Island. I think of Paul Winter, whose songs celebrate the whale, the porpoise, and the Australian aborigines. I think of Susan Griffin, whose prose calls us back to things and more things, all bound together in an erotic continuum of being. Yet few among us are poets, essayists, and songsters. I cannot envision a future that depends solely on their energy. For it is all of us who are speaking humans, embedded in the text that is culture (and more particularly in that narrative tradition that is MAN), afflicting wild creatures and lands, driving them into oblivion.

Consider the word *environment* and all its permutations in contemporary *environmentalism*. Other words also lend themselves to such inquiry, such as *ecosystem*[24] and *resource.*[25] A reflexive study reveals that such words, bound up in the ideology of environmentalism, serve to distance MAN from earth. Deep ecologists are sometimes called "radical environmentalists." We find environmental science, environmental geology, environmental policy, environmental studies, environmental economics, environmental activism, environmental organizations, environmental movement, environmental protection, and, yes, even environmental ethics (with its *theories* of rights, values, duties, and so on). And there are environmental protection agencies, environmental impact statements, environmental information systems, environmental hazards, and even environmental perception. There are entire dictionaries and even encyclopedias of the environment.

Who among us can be opposed to the word family that environmentalism represents? Yet could it be that environment is more symptom of than solution for anthropogenic mass extinction? Is the word *environment* mendacious? Environmentalism perhaps conceals the possibility of actually conserving life on earth. Environmentalism perhaps allows us to stand apart from a mute world of matter in motion, now become standing reserve for human appropriation. The problem, David Quammen suggests, is not that environmentalism "in its essence is perverted. It's just an understandable campaign of self-interest, by our species, with potentially dire implications for the world at large. What does seem perverted is confusing environmentalism with conservation."[26]

Contemporary usages trace from the early nineteenth century (ca. 1825); that usage itself derived from older roots (ca. 1300) in Middle English *(envirounen)* and Old French *(environner)*. These in turn were derived from *viron,* to circle, as in *environ,* that is, to circle around. We find Chaucer, for example, using the term *environing* to mean circling around or surrounding. The Oxford English Dictionary lists the action of environing and the collection of entities or things environing as the two primary meanings of environment—which I supposed is all well and good.

Yet consider that a leading contemporary dictionary of scientific literacy defines environment as "everything that makes up *our* surroundings. In the physical world, the term means the global or local conditions affecting *our* health and well being."[27] This definition seems remarkably consistent with the Cartesian–Baconian notion that we are the owners, that is, masters and possessors, of nature. Another contemporary dictionary defines environment as either "the combination of external conditions influencing the lives of individual organisms" or "the internal conditions, primarily chemical, that control the well-being of the individual plant or animal."[28] In the first definition the earth is something to which humankind is related via external causation (action at a distance); there is no idea here that we are somehow in nature, a part of nature. In the second definition nature has been reduced to chemistry, as if our bodily organism itself is not both influencing and influenced by the earth.

Is "environment," Quammen asks, "a bad label for a good idea or a good label for a bad one?" Has scientific literacy helped convert, through some strange kind of linguistic alchemy, wild nature into environment, into a thing that becomes only an object of study, a scientific study that reveals nature's laws and principles of operation so that we might have causal control? The word, Quammen contends, "entails a presumption that humanity is the star of a one-character drama around which everything else is just scenery and proscenium."[29] The world becomes a stage that we manage and direct to our exclusively human purposes. Can it be that our ecological abnormality is tied to environment? Is environment a projection of MAN?

How is earth-talk to change things, to turn reform environmentalists into conservationists who realize that the human fate is deeply entwined with the fate of others, teaching that conservation is something more than sustainable development, clean air, and clean water? After all, it is the environmentalists who want political power, who seek to manage the planet and promote sustainable development. Better perhaps for those who feel an affinity with wilderness to seek solace in alpine meadows, to lose ourselves in the dark lanes of the summertime Milky Way, to wander with migrating elk in the fall, to vibrate with the leaves of the aspen.

Consider the possibility that inherent in language is the energy of the cosmos, the power of the wind and the sea, the vitality of animals, and the persistence of plants. Could it be that the same cosmic forces that move the sun and the earth also move mortal flesh? The space prepared for us by our ancestors, the scribes who created the Greek alphabet, the philosophers who first used it, the inventors who created movable type, enframes us. And yet we can pass through this second world, this artifice, to the other side. For concealed by the text is a primordial truth: the awesome birth of vociferation, the explosive origin of speech and sociality.

The gains of literacy are manifest; there is no way back per se. I myself am a bibliophile and an intellectual. I have made and am making contributions to encyclopedias of environmentalism. Good, honest, linear, logical, analytical work. Euroman at the word processor. Making lists. Categorizing. Understanding. But writing, as Walter J. Ong notes, "is the seedbed of irony, and the longer the writing (and print) tradition endures, the heavier the ironic growth becomes."[30] Language speaks! Perhaps philosophy, Merleau-Ponty suggests, "consists in restoring a power to signify, a birth of meaning, or a wild meaning."[31] Insofar as philosophy retains any meaning, it must seek to reestablish connection with the biophysical processes that engendered our species and yet sustain civilization. Merleau-Ponty counsels us to think of the awesome birth of vociferation. We are flesh of the world, speaking flesh bound with the reality of evolutionary process, both cultural and natural.

Imagine that the space in which we dwell is a story (a complex story to be sure), a powerful cultural narrative communicated among us all. The story of MAN. Euroman, in control of the environment. Euroman, who will manage planet Earth. Euroman, who will initiate the process of sustainable development.

Could it be that the story of Euroman is pathological? Most Westerners are living, as Thoreau claims, lives of quiet desperation. Primal peoples have been displaced from their lands and deprived of their culture by Euroman's relentless advance, always in the name of civilization. Wild nature has been afflicted by Euroman, since nature is no more than environment, than resource.

We are caught between a failed story, the past that yet overdetermines us, and another that is powerless to be born, the future that is not yet but might come to be. I find the new story—that we are of and about earth—in gestation all around me. I hear again and again the earth-talk of ecologists and mountaineers, activists and defenders of wildlife, organizers and wildlife caretakers, who sing and celebrate the earth, who lay down the illusion of the *ego cogito* and accept themselves as flesh of earth.

Surely, now, I've gone too far, if not committed religious apostasy. But I am serious. And speaking literally, not poetically. The road from here—that is, from a world mired in an ever-worsening ecocrisis that verges on ecocatastrophe—to there—that is, to a culture built on the practice of the wild—is marked by the signposts of language. There is no other way for *Homo narrans*. To change culture one must reconsider language, radically reconsider language.

Earth-talk, I am claiming, is a way in which the more-than-human might again give voice and be heard. Earth-talk might reestablish communication with the more-than-human; on that basis we might speak authentically of our common future—a common future that goes far beyond reform environmentalism. Today the biophysical world has become environment, little more than raw material that reform environmentalism will manage as *Homo sapiens* pursues the goal of sustainable development. Earth-talk reminds us of the phenomenon of displacement. Earth-talk is the *via media* toward an ecology of language. But what is the ecology of language?

In the first place, it is a problem. For through literacy we have been displaced from the green world from which we came. As Loren Eiseley notes,

"Linguists have a word for the power of language: displacement. It is the way by which [the human species] came to survive in nature. It is also the method by which [humankind] created and entered [our] second world," the cultural realm that conceals our origins, our home, from us.[32] But if it is through language generally, and through literacy more particularly, that we have been alienated from the first world, then it is also through language that we can return.

That project is now under way. But it is monumentally difficult. Gregory Bateson writes, "The problem of how to transmit our ecological reasoning to those whom we wish to influence in what seems to us to be an ecologically 'good' direction is itself an ecological problem. We are not outside the ecology for which we plan—we are always and inevitably a part of it."[33] Our talk, insofar as it speaks of earth and the conservation of biodiversity, necessarily involves the ecology of language—language that reconnects the human animal (*zôon logon echon*) with the more-than-human. Around us we find thousands of individuals who speak earth-talk in dozens of new idioms, that is, in a variety of ways that begin to interface all living things into a single community of life.

A variety of ecologically minded thinkers are working in the fertile field of the ecology of language (though none of them, to my knowledge, characterizes their work so). Consider the work of Jim Cheney.[34] By Cheney's account there are no general solutions for ecological dysfunctions, but only solutions in relation to specific ecosystems in which human beings have effectively taken root. Sustained living in place, he claims, gives humans the opportunity to discover the fundamental rhythms and pace, the structures and dynamics, of particular ecosystems. Bioregional narratives reflect human placement in nature and articulate local knowledge of the interrelations between the human and the more-than-human. They also enable storied residence, where the transcendental subject and the notion that language as vehicle for expression of human intentionality is challenged by individuals who come to self-consciousness in place; such individuals realize that the association of person with place is more intimate than MAN realizes.

Storied residence, an idea that Cheney takes from Holmes Rolston's work, offers a new way of being in the world, an alternative to reform environmentalism that reduces the diversity and particularity of local places to one universal mechanism: nature as matter in motion. In Cheney's view, bioregional narrative does not reduce the biophysical world to language but rather incorporates that world into ecologically nuanced human discourse. Within that frame the living world can disclose itself, that is, speak through human beings who have taken up being in the world as plain members rather than as absolute rulers of the land community (as Leopold might put the point). Christopher Manes's work can also be read as extending the ecology of language. Manes looks explicitly at the language that posits MAN as speaking-thinking subject apart from and in control of the earth. "The language we speak, today, the idiom of the Renaissance and Enlightenment humanism, veils the processes of nature with its own cultural obsessions, directionalities, and motifs that have no analogues in the natural world."[35] Manes calls systematic environmental ethics into question, since it is an abstraction from the life-world of immediate human experience, the gesture of MAN who is in control of the planet. Systematic philosophy is characterized by Manes as the discourse of reason that reflects the disembodied ego inherent in modern language, and the consequence is the silencing of nature's voice. Manes notes that primary oral cultures have allowed the flora and the fauna "to speak." He valorizes deep ecology as perhaps the most linguistically open form of ecophilosophical discourse, but he carefully qualifies this contention. Humans need to speak a language, Manes argues, that cultivates a sense of ontological humility, reconnects human projects with the larger earth community, and moves us beyond our preoccupation with MAN.

A third example in the ecology of language is the work of David Abram.[36] Working outward toward the world and the reality of lived experience from Merleau-Ponty's posthumously published *The Visible and the Invisible,* Abram attempts to disclose a transcendental signified embedded in the narratives of environmental philosophers: namely, the disembodied thinking subject, the *ego cogito,* indeed, the environmental philosopher who is held apart from and above the world by his reason. Who is this ego that thinks? What is it? The systematic philosopher does not entertain such questions, for his language is assumed to be transparent, a virtual mirror of reality. Abram also explores the phenomenology of perceptual experience, exploring a bodily discourse concealed beneath the customary forms of linguistic expression. We are, Abram insists, embedded as organisms in the awesome mystery of the corporeal world. Language itself reflects this primordial reality, the reality of lived experience and participation of the body in the world.

No philosophical writer today can claim to be anything but an ironist. There is no quest for Platonic universals. No hope of a Hegelian absolute. No dream of Husserlian *wesen.* No plea for a unified field theory. Narratives, whether cultural or individual, are just that.

Accepting contingency does not mean, however, that polysemy is infinite.

We storytelling culture-dwellers are embedded in traditions or paradigms that constrain discourse, that direct it. Earth-talk should be free of the metaphysical presuppositions, the binary oppositions such as spirit and matter, man and woman, the wild and the civil, that run through the language of environmentalism. Wild nature and the plant and animal others become genuine conversation partners. Language is recontextualized as a *via media* between humankind and the more-than-human, an open, fluctuating system of signs upon which our survival as a species likely depends. From such a position, inside language, we reestablish ourselves as natural animals while remaining distinctively human.

And is it not the case that the ecology of language reconnects us to the earth, to the beauty of things, to the mystery of life? It is not by accident that I have found the ecology of language at work in Winter, Griffin, and Snyder —in song, prose, and poetry. This is the domain of what traditionally is called the arts and humanities (though that term, too, is laden with pitfalls). From inside language it may be possible to reconceptualize environmentalism, much as R. J. Johnston suggests.[37]

Try thinking of environmentalism as a struggle between two social elites, one socially dominant and in control, and the other a minority that questions the legitimacy of the status quo, including (in my case) environmentalism itself. The "technical intelligentsia," as Johnston calls them, run the agencies, research laboratories, and allied university programs that control the environmental planning and technological development process. In academic jargon, they are at the center of society. Their goal is either to overcome or to ameliorate the ecopathologies engendered by the massively growing human population and the dominant mode of production. But there is to be no basic change in the trajectory of history: managing earth and sustainable development are the dominant gestures and *efficiency* is the bureaucratic watchword.

The socially dominant elite confines ecology to environmentalism. It speaks only of point and nonpoint sources of pollution, toxicants and risk assessment, Superfund cleanups, sustainable development, and so on ad infinitum. There is no thought here that ecology in any way speaks to us as human beings, to the very conception that we have of ourselves, for this elite is embedded in a discourse of power, in a conceptualized hierarchy that places MAN the rational animal in control of the ecomachine. Nature is standing-reserve, subject to appropriation for MAN's economic purposes. In short: Environment.

The "humanistic intelligentsia," as Johnston calls the minority, are on the margin, largely excluded from the agencies and institutes that seek to address ecocrisis. They believe that the dominant mode of human reproduction and production is itself the problem. Thus, they seek not to rationalize the continuing destruction of earth under the banner of sustainable development, but instead attempt to recontextualize the issues themselves.

How peculiar and quaint those who make the noises of earth-talk seem to the technocratic elite! What could be more irrelevant to saving the planet than the land community, land ethics, and thinking of ourselves as flesh of earth? The socially legitimated and institutionalized ideology, as Johnston notes, "is cornucopian, with a strong belief in human ability to dominate nature, through science, technology and their application via the marketplace, and with the necessary support of the state."[38] The paradox of environmentalism suggests that the situation is otherwise: we are not in control but simply along for the ride, mired in a failed story and as yet incapable of creating a viable alternative.

I have new reason for hope these days, engendered partly through my encounters with the writing of Wilson—texts I've started to read in new ways. My respect for Wilson's stature as a conservationist continues to grow as I more fully grasp the fact that the ecology of language is running wild in his many books, coloring them with nuances that celebrate the profusion and glory of life. The path he takes is like a trail through an alpine forest: it is exhilarating. This way, this road, I've come to believe, is the path to a viable future for life on earth. Wilson tells us that we stand at a rare juncture in time when an innate human propensity to affiliate with life is aligned with reason. "To explore and affiliate with life," he writes, "is a deep and complicated process in mental development. To an extent still undervalued in philosophy, religion and science, our existence depends on this propensity, our spirit is woven from it, and hope rises on its currents."[39]

Who is it, I asked earlier, that speaks? Is it not the case that we are of and about earth? That our language, as a poet once said, "is everything, since it is the voice of no one, since it is the very voice of the things, the waves, and the forests."[40] Communication. Community. Earth-talk.

Notes

1. See Arne Naess, "Intrinsic Value: Will the Defenders of Nature Please Rise?" in *Conservation Biology: The Science of Scarcity and Diversity*, ed. Michael E. Soule (Sunderland, Mass.: Sinauer, 1986), 512.

2. Edward O. Wilson, *The Diversity of Life* (Cambridge: Harvard University Press, 1992), 32.

3. Wilson, *Diversity of Life*, 351.

4. Edward O. Wilson, "Threats to Biodiversity," *Scientific American* 261 (September 1989): 108-16.

5. The challenges conservation faces are far more complicated than biodiversity, if for no other reason than that virtually all aspects of ecocrisis interact. See, for example, *Global Warming and Biological Diversity*, ed. Robert L. Peters and Thomas E. Lovejoy (New Haven: Yale University Press, 1992); this text contains essays that explore the pervasive interconnections between climate change and the extinction of species.

6. See Max Oelschlaeger, "The Idea of Wilderness as a Deep Ecological Ethic," paper presented at the Fifth World Wilderness Conference, Tromsø, Norway, September 1993.

7. Vandana Shiva, *Staying Alive: Women, Ecology, and Development in India* (London: Zed, 1989).

8. See World Commission on Environment and Development, *Our Common Future* (New York: Oxford University Press, 1987).

9. See Manfred Stanley, *The Technological Conscience: Survival and Dignity in an Age of Expertise* (Chicago: University of Chicago Press, 1978).

10. Paul Kennedy, *Preparing for the Twenty-first Century* (New York: Random House, 1993), 53.

11. Marshall McLuhan, *The Gutenberg Galaxy: The Making of Typographic Man* (Toronto: University of Toronto Press, 1962), is an able introduction.

12. See Paul Shepard, *The Tender Carnivore and the Sacred Game* (New York: Charles Scribner's Sons, 1973) and *Nature and Madness* (San Francisco: Sierra Club Books, 1982).

13. See Max Oelschlaeger, "Wilderness, Civilization, and Language," in *The Wilderness Condition: Essays on Environment and Civilization*, ed. Max Oelschlaeger (San Francisco: Sierra Club Books, 1992), where I argue the case in detail.

14. Marjorie Grene, "The Paradoxes of Historicity," in *Hermeneutics and Modern Philosophy*, ed. Brice R. Wachterhauser (Albany: State University of New York Press, 1986), 182.

15. Walter J. Ong, *Orality and Literacy: The Technologizing of the Word* (London and New York: Routledge, 1982), remains the finest introduction to the subject.

16. Calvin Luther Martin, *In the Spirit of the Earth: Rethinking History and Time* (Baltimore and London: Johns Hopkins University Press, 1992), 18.

17. Carolyn Merchant, *Ecological Revolutions: Nature, Gender, and Science in New England* (Chapel Hill and London: University of North Carolina Press, 1989), 29.

18. See Paul Shepard, "A Post-Historic Primitivism," in Oelschlaeger, *Wilderness Condition*.

19. Neil Evernden, *The Natural Alien: Humankind and Environment* (Toronto: University of Toronto Press, 1985), 124.

20. John Firor, *The Changing Atmosphere: A Global Challenge* (New Haven: Yale University Press, 1990), 103.

21. Wilson, *Diversity of Life*, 272.

22. See Christopher Manes, "Nature and Silence," *Environmental Ethics* 14 (1992): 339-50.

23. See Julian Burger, *The Gaia Atlas of First Peoples: A Future for the Indigenous World* (New York: Anchor Books, 1990); Shiva, *Staying Alive*; and Ramachandra Guha, "Radical American Environmentalism and Wilderness Preservation: A Third World Critique," *Environmental Ethics* 11 (1989): 71-83.

24. See Leslie A. Real and James H. Brown, eds., *Foundations of Ecology: Classic Papers with Commentaries* (Chicago: University of Chicago Press, 1991).

25. See John W. Bennett, *The Ecological Transition: Cultural Anthropology and Human Adaptation* (New York: Pergamon Press, 1976).

26. David Quammen, "Dirty Word, Clean Place," *Outside* 8 (1991): 26.

27. Richard P. Brennan, *Dictionary of Scientific Literacy* (New York: John Wiley & Sons, 1992), 98, my emphasis.

56 . Max Oelschlaeger

28. Gareth Jones et al., *The Harpercollins Dictionary of Environmental Science* (New York: HarperCollins, 1992), 135-36.

29. Quammen, "Dirty Word," 25.

30. Ong, *Orality and Literacy*, 103.

31. Maurice Merleau-Ponty, *The Visible and the Invisible*, ed. Claude Lefort, trans. Alphonso Lingis (Evanston: Northwestern University Press, 1968), 155.

32. Loren Eiseley, *The Invisible Pyramid* (New York: Charles Scribner's Sons, 1970), 144. On language and survival see Edward O. Wilson, *Sociobiology: The Abridged Edition* (Cambridge, Mass.: Harvard University Press, 1980), and Richard Dawkins, *The Selfish Gene* (New York: Oxford University Press, 1978).

33. Gregory Bateson, *Steps to an Ecology of Mind* (New York: Ballantine Books, 1972), 504.

34. Jim Cheney, "Postmodern Environmental Ethics: Ethics as Bioregional Narrative," *Environmental Ethics* 11 (1989): 117-34.

35. Manes, "Nature and Silence," 339.

36. See David Abram, *Environmental Ethics* 10 (1988): 101-20; also see David Abram, *The Earthly Body of the Mind: Animism, Language, and the Ecology of Sensory Experience* (New York: Pantheon, 1996).

37. R. J. Johnston, *Environmental Problems: Nature, Economy and State* (London: Belhaven Press, 1989), 185-86.

38. Ibid., 199.

39. Edward O. Wilson, *Biophilia* (Cambridge, Mass.: Harvard University Press, 1984), 1.

40. Merleau-Ponty, *The Visible and the Invisible*, 155.

11 . CROSS-CULTURAL WILD

4 . Pancultural Wilderness

Marvin Henberg

> We need wilderness preserved—as much of it as is still left, and as
> many kinds—because it was the challenge against which our
> character as a people was formed.
>
> *Wallace Stegner[1]*

Wallace Stegner's rationale for wilderness preservation addresses the mainstream of North American culture. His words suggest that rationales for
wilderness preservation may be specific to one culture or another. Perhaps,
as thoughtful people from many different cultures suggest, there is no such
thing as a single view of wilderness that will do for all cultures.

Lacking such a pancultural view of wilderness, the case for preservation
must proceed internationally on an ad hoc basis. Cultures that discover internal reasons for valuing wilderness will sprout wilderness preservation
movements. Other cultures will not. We seem to get both ourselves and
those we might be tempted to criticize off the hook so long as we accept the
relativistic tenet that no culture may legitimately use its standards to judge
another.

Deeper consideration of the preceding affirmation of cultural relativism,
however, shows that it begs the question. It may be wrong to judge one culture by standards specific to another, but that judgment does not address the
possibility of pancultural criticism. My goal is to argue that cultures can and
do negotiate mutually agreeable, pancultural ideals.

This undertaking has four stages. The first is to place the wilderness
movement in the context of other pancultural enterprises. The second stage
is to argue, at a conceptual level, against two varieties of epistemological relativism that would, if valid, preclude all possibility of a pancultural idea of
wilderness. The third is to consider the challenge posed by cultures that lack
the wilderness concept altogether. The fourth and final stage is to consider
what I believe to be the real difficulty arising from a clash of cultures.

Cultural differences are to be expected. They present a problem only
when reinforced by attitudes characteristic of a specifically *authoritarian*
culture. The idea of wilderness as a place of freedom for nonhuman animals
and plants has substantial prospect of acceptance across many diverse cultures as long as it is not blocked by authoritarian casts of mind.

Pancultural Ideas in Science and Politics

To fix our mark, let us consider an unproblematic example of a pancultural idea—one drawn, as so many are, from natural science. Take the view that the Earth is several orders of magnitude more ancient than the genealogies in Scripture. Among the international community of scientists, the idea of an ancient age for our planet is accepted by Hindu and Christian and Moslem alike, by Argentine and Russian and Korean, by ethnic Kurd and ethnic Serb. To the extent that scientists drawn from these diverse peoples know and care about the evidence, they accept the geological proposition of an earth six billion or more years old. It matters not a whit that the idea's original proponent, James Hutton, was a Scot. It would be absurd to argue that the idea is culturally relative and that only Scots could find it acceptable.

By analogy, if the idea of wilderness were an idea of science, it would be equally absurd to argue that because the original proponents of wilderness preservation—George Catlin to John Muir to Robert Marshall—were citizens of the United States, the idea is acceptable only there. Of course, the rub is that the idea of wilderness seems to be anything but an idea of science. It is rather a political term used in persuasive writing for or against a particular kind of land preservation.

In this respect, wilderness is more like human rights than like a concept from natural science. The idea of human rights is moral and political, an idea of persuasive art whose success (especially in the wake of the Holocaust) makes it a strong candidate for panculturalism. The concept of indefeasible human rights is considerably more than a mere invention. It addresses human aspirations in a way that advances its own adoption. It has a legitimate place in the pantheon of human ideals, for it speaks across cultures to people as rational beings, capable of creating a better future.

As a political idea, wilderness is even younger than the relatively youthful concept of human rights. Still, it has a history, and we would do well to reflect on it. The fourth World Wilderness Congress (WWC) in 1987 labored mightily to address deficiencies in the idea as articulated legislatively in the United States, particularly in the 1964 Wilderness Act.

Historically, Congress was privileged in two respects when it came to designating wilderness lands. First, compared to much of the world, the United States faced an embarrassment of riches. No matter how greatly wilderness advocates bemoaned the shrinking of roadless lands, the country possessed—relative to Europe, Japan, and many other developed regions— enormous potential for effective preservation. The United States could afford the presumption that lands considered for wilderness had to be primordial, with a continuous history of apparent noninterference by human beings. The fourth WWC rightly amended the American understanding so that restored natural lands, though lacking a continuous history of human noninterference, might qualify.

Second, legislators in the United States were heirs to a history of dispossessing Native Americans of their lands, rights, and cultures. Accordingly, Native Americans were ignored in the 1964 Wilderness Act. I suspect this sad tale is much in the minds of those who charge that the idea of wilderness is a culturally specific invention of the white, European-influenced civilization that came to dominate North America. To support the idea of wilderness, these critics fear, tacitly supports the genocide and dispossession of Native Americans. Analogously, to support wilderness preservation internationally tacitly supports the genocide and dispossession of other indigenous peoples.

The better part of wisdom is to amend our definition to be sensitive to the rights and interests of indigenous peoples inhabiting or otherwise utilizing proposed wilderness reserves. This amendment was undertaken by the fourth WWC:

> Wilderness is an enduring natural area, legislatively protected and of sufficient size to protect the pristine natural elements which may serve physical and spiritual well-being. It is an area where little or no persistent evidence of human intrusion is permitted, so that natural process may begin to evolve.[2]

This definition countenances the presence of indigenous peoples (say, hunter-gatherers) whose traditional cultures go lightly upon the land, leaving few intrusive signs. It also endorses relocation of indigenous peoples such that lands formerly under their sway may return to nature. Forced dispossession, as in the United States, is by no means the sole route to achieving a relocation of human population in the interest of wilderness.

It is important to note that the fourth WWC definition is the product of discourse and debate across cultures. The spread of ideas from one culture to another is not always imperialistic, with a dominant culture always forcing its ideas on others. Such a belief sometimes leads to a mistaken embrace of cultural relativism as a supposed antidote. By contrast, once it is clear that pancultural ideas can spread on the basis of mutual consent, the charm of cultural relativism is diminished. We are encouraged to hope that a shared idea of wilderness will someday enjoy universal assent, at least among nonauthoritarian cultures.

To the extent that hope fuels activism, wilderness defenders should behave as if their favored ideal can and will become pancultural. Like all great ideas nurtured from seed, the idea of wilderness must speak to many different understandings of nature and reality, and it must do so powerfully over the next fifty years. Otherwise it may well prove an oddity, a curiosity of twentieth-century cultural conditions in North America. In the global village of tomorrow, wilderness will have little currency if it fails to be negotiated and voluntarily embraced across cultures.

The Challenge of Epistemological Relativism

The fourth WWC definition will not satisfy those who believe on epistemological as opposed to historical or cultural grounds that the idea of wilderness is relative.[3] Contemporary philosophy provides two natural homes for epistemological relativism. The first and more radical lies in the body of deconstructive literary theory associated with Jacques Derrida.[4] According to this body of theory, the meaning of every concept potentially differs between speaker (or writer) and listener (or reader). There are no privileged and timeless meanings, but rather a cacophony of meanings, all dependent on the sex, ethnicity, family background, education, and power status of the person interpreting the concept. According to deconstruction, the concept of wilderness lacks a fixed meaning because *all* concepts lack fixed meaning.

The second, more modest, natural home for contemporary epistemological relativism lies in the views of Anglo-American philosophers such as W. V. O. Quine, Nelson Goodman, and Richard Rorty.[5] These philosophers stress the link between any particular concept's meaning and the meanings of all other concepts within the symbolic system (usually a natural language) employed to express the concept. Their views raise doubts, not about fixed meaning per se, but about the feasibility of determining it in any given case.

Debate between epistemological relativism and its contrary, epistemological absolutism, is perennial—one of the stock trades of philosophy. I shall not settle this debate here. At the highest level of generality, there may be no resolution to the dispute, for absolutism (the defense of fixed and timeless meanings for concepts) and relativism (the insistence that meanings differ among persons, eras, or cultures) each have more the trappings of faith than of rational conviction.

At another level, however, the two strains of contemporary relativism—that of deconstruction and that of Quine et al.—can be answered. There are pragmatic reasons for rejecting both. Whenever in discourse people use

common terms and seem to agree (even if it is only to agree that they disagree), then we should assume, skeptical arguments to the contrary, that
they *do* understand one another. The commonalities of human existence—
birth, death, and making a living—seem a sufficient practical basis from
which to declare that at least some concepts are universal. This impression is
reinforced by considerations from evolutionary naturalism: our common
human ancestors were shaped by a distinct set of environmentally selective
forces, and it would be astounding to find that these were of no importance
when it comes to providing a common content for basic human concepts.

Accordingly, the most radical strain of contemporary epistemological relativism—that found in much of deconstructive literary theory—can be
pragmatically ignored, if not disproved. The second strain, however, raises
concerns of a different order. The idea that a concept's meaning is relative to
particular language systems (rather than to particular persons) presents a
novel challenge to the possibility of a pancultural view of any concept,
wilderness included.

Take Quine's thesis of the indeterminacy of translation. The thesis goes as
follows: translations among different signifying systems (natural languages
among them) may be globally reliable, but will always be locally unreli-
able—that is, indeterminate in any specific instance. A concept's meaning is
determined in part by its relation to allied concepts in its linguistic system.
Even if we presume that all rational human beings' prelinguistic perceptions
of reality are identical, different languages may apportion these common
perceptions differently among the complex set of allied concepts necessary
for adequate description. To take Quine's famous example, an English
speaker hears the word *gavagai* as a speaker of an unknown native language
points to a rabbit bounding across a field. The word *gavagai* may seem to
mean "rabbit," when the native tongue includes the meanings "swift running
beast" or "something tasty for lunch."[6] Because human beings lack nonlin-
guistic means of pointing to this difference, it will always be opaque to translation. The native concepts closely allied to *gavagai* will themselves, of necessity, be sufficiently different from the English concepts allied to *rabbit* that
they too will suffer indeterminacy in translation.

The problem is conceptually akin to the dilemma in which we find ourselves when under the sway of Cartesian doubt. Each time anyone tries to
portray the grounds for universal doubt by giving examples of particular
doubts, that person cannot make any particular doubt plausible without assuming a whole host of other things to be true. Doubt, I believe, is conceptually relative to things not doubted in the same way a concept in a language

system derives its meaning from its allied concepts. If we change the things not doubted that give us grounds to doubt, we affect the particular thing we doubt, making it either more or less dubious than before. If we change the allied conceptual signifiers that give a particular concept at least part of its meaning, we correspondingly change that meaning.

According to Quine, translating a concept between languages remains indeterminate precisely because its crucial allied concepts are themselves objects of translation. Alleged conceptual identities between two language systems are either real identities or the result of an undetected (and in principle, undetectable) mistranslation of allied concepts. Our problem, the human problem, is that we cannot tell one case from the other.

But is this such a big problem? If a translation works—if two people communicate in such a way that a difference in meaning is undetected—why worry? Why not say simply that over a broad spectrum of possible uses, the two speakers employ the term in the same way? For pragmatic purposes meaning is use. Like indeterminacy in quantum mechanics, indeterminacy in translation is a microscale problem. At the macroscale of human practice and interaction, the indeterminacy can be ignored. If the difference in meaning begins to make a difference in practice, it will be noticed. Someone will cry foul, and the indeterminacy will be resolved.

I would argue, then, that both varieties of modern epistemological relativism are insufficient to establish that the idea of wilderness lacks pancultural scope and influence. According to the relativism of deconstructionism, wilderness is no different from any other concept. The thesis that a concept is culture bound becomes trivial. The idea of a city, of a nation, of truth, of justice . . . all are cultural artifacts. Therefore, to claim that any one of them is a cultural artifact is to say something analytic, something on the order of "a word is a word" or "a material object is something that persists through time." According to the relativism of Quine, we may be forced into skepticism about the accuracy of translating terms from one symbolic system to another, but that is all. On a global scale, the translations are reliable so long as persons using different language systems can interact and understand one another.

Concept Lacking Altogether: A Deeper Challenge

When nonphilosophers proclaim the idea of wilderness to be relative to culture, they likely mean something other than that every concept is relative to culture or that a concept, taken individually, has an indeterminate translation. Irrespective of any thesis about the relativity of all terms, some terms may be specific to culture in an especially strong way. We have more to worry about than nuances in translating roughly synonymous terms between different languages. Perplexing difficulties arise when we encounter a term that is meaningful in one language system but has no conceptual analog whatsoever in other systems. It isn't a matter of dividing up allied concepts or parsing related words a bit differently. Instead, the concept simply isn't there, isn't available for *any* translation.

Under this light, the idea of wilderness might be culture bound in the sense that some cultures possess the idea whereas others lack it. Take the following famous passage from Standing Bear, an Oglala Sioux:

> We did not think of the great open plains, the beautiful rolling hills, and the winding streams with their tangled growth as "wild." Only to the white man was nature a "wilderness" land, only to him was the land "infested" with "wild" animals and "savage" people. To us it was tame.[7]

The Oglala Sioux did not have a word corresponding to the English word *wilderness*. Nor do many other languages, either because their cultures no longer possess lands unmodified by human action, or because their cultures direct them to live in a relatively ecofriendly manner, as with many Native American tribes prior to contact with whites.

Such cultural relativity, however, does not imply that the idea of wilderness is pure invention, a fiction. If we believed this thesis generally about concepts with historical roots in one specific culture, it is difficult to imagine why people would, as they so frequently do, seek out alien ideas for incorporation into their own worldviews.

Prospects are good, I believe, for the idea of wilderness to imitate the idea of human rights, winning a place in the pantheon of legitimate ideals. The test for wilderness as a pancultural ideal lies not so much in the past as in the future. The relevant question is not whether Standing Bear had an idea of wilderness, but whether a representative of his sensibility, knowing what human beings now know about civilization's omnivorous appetite for land and resources, would *today* understand the idea. By this test, I think, we would find Standing Bear an advocate of wilderness, seeing it as the most viable option for preserving that which is essential to the old ways, his people's land. This conclusion is more than ad hoc hypothesizing on my part, for at least two Native American tribes, the Yakimas and Flatheads, have designated portions of tribal land as wilderness.

A relativist may object that, Native Americans aside, the preservationist

ethic embodied in today's wilderness management is astounding, even incoherent, to other traditional, nonindustrial peoples. Ideas such as letting
alone, keeping wild animals simply to be seen, and not chopping firewood
from the forest make no sense to many peoples of the globe.[8]

I have two observations with respect to this objection. First, while there
are many things that members of traditional cultures fail to comprehend,
this fact alone is a far cry from establishing that they are *incapable* of comprehension. It is difficult, for instance, to imagine explaining a lunar lander
to a person from a culture lacking sophisticated technology. If a practice or
use for a concept does not exist within a culture, its members will not have
it—that much is uncontroversial. But this lack does not prove the concept to
be culturally relative in any deep and permanent sense. Will no one from the
alien culture ever understand the concept? It is difficult to have faith in education, travel, or cross-cultural exchange if one holds such a strong view of
cultural relativity.

My second observation builds on the important difference between a person's lacking a concept altogether and having the concept but failing to value
it as others do. Nothing we have considered so far speaks to the attitudes, pro
or con, that people from different cultures may have toward a practice or
idea; yet this realm of value is where most of the controversy lies. It is also
where some people find the idea of cultural relativism to carry most weight.
Even here, however, there is a tendency to exaggerate the differences among
us. Take, again, the spread of concepts in modern science. In virtually every
field, meteorology to ethology to physics, the scientific community is international. Standards and practices for investigating and validating hypothe-
ses are largely agreed on, even if results and interpretations are not. International congresses, exchanges with laboratories in other countries, education
abroad—all are the norm in science. This demonstrates that ideas of science
are both pancultural and positively valued: the coming together of scientists
from many cultures in common pursuit of understanding indicates near
unanimity in the value judgment, "This enterprise is important."

The Problem of Authoritarian Culture

Finally, I suggest that problems arising from clashes between cultures,
though real and important, are more likely to be masked than illuminated
by the embrace of cultural relativism. From the fact that peoples differ in
their practices and belief systems, we may infer that pancultural understanding is difficult, but we should not infer that it is impossible. Ample tes-

timony from history and psychology suggests that human beings crave to communicate and to forge mutual understanding.

On inspection, an idea's failure to communicate across cultures is often a function either of its sheer arbitrariness or of an authoritarian cast of mind. Matters of taste and convention are examples in the first category. Cultures sometimes select for and prescribe behavior for reasons lost to contemporary understanding. Here tradition becomes the sole justification, and where traditions clash between cultures, neither side may be motivated to efface the difference. Even so, both sides can accept the difference, which may in turn lead to mutual recognition of the right of others to maintain differing tastes and preferences.

The problem becomes intractable only when beliefs and practices are combined with authoritarianism. To hold a belief is to hold its propositional content to be true, whether the believer holds the conviction as a member of a particular culture or as an individual. At the same time, individuals in every culture have had the experience of believing something that turned out to be false. The danger lies not in the dogmatism of belief per se, but in the cast of mind that opens or closes a person to possible revision of belief. It is usual for members of a culture to insist that their views are true; without such a claim, it is hard to imagine how else they would maintain identity. A breakdown in understanding becomes permanent only when members of a culture take the further step of also claiming a monopoly on approved methods of confirming truth. Such is the tenor of an authoritarian culture: its members insist they have privileged access to the relevant means of proof.

We are now in position to understand the real challenge of cultural clash to the idea of wilderness. The intractable problem arises not from mere differences in belief, but from a particular reinforcement of belief on at least one side. Danger arises when any cultural group claims its version of reality to be not only true, but self-referentially validating as well. This claim flies in the face of science as it is understood and practiced worldwide. Science is antiauthoritarian when it comes to argument over truth and falsity. The humblest of intellects is welcome, in principle, to propose and test any scientific hypothesis. Claims are adjudicated on grounds that are both public and intersubjective, closed to none.

The real problem, then, does not arise from just any conflict of cultures, but from a clash in which one or more is an authoritarian culture. Doubtless some defenders of the original North American concept of wilderness have fallen into this trap, promoting their idea as true by privileged means. We should resist this and every other pronouncement of authoritarian culture, but not to the extreme of denying the very possibility of an emergent pancultural view of wilderness. This position is itself authoritarian.

A critic may at this point observe that some cultures do not value modern science. To the contrary, some cultures view repudiating science as imperative in preserving their traditional ways. If such is their choice, the reasoning goes, we should not interfere—an obligation we supposedly observe in our embrace of culture relativism.

But are we really cultural relativists in deriving such a conclusion? Notice how the preceding invocation of relativism contains an appeal to the moral values of freedom and tolerance. The insistence is that traditional peoples ought to be free to pursue their own ways so long as they do not threaten others. The "ought" in this conclusion expresses a value judgment. Either this judgment is valid for all cultures or it is not. Let us trace out the logic of these alternatives.

Suppose the value judgment "Traditional peoples ought to be free to pursue their own ways so long as they do not threaten others" fails to be valid for all cultures. Such a view rejects freedom and tolerance as universal goods. Any person who rejects tolerance as a universal good but who still wishes to pitch for noninterference in the ways of traditional, indigenous cultures is in a peculiar situation. That person's opponent—the cultural chauvinist—says we have the might and therefore the right to push these people aside, utilizing the resources of their home territory as we see fit. The one who would defend the indigenous culture from a relativist's perspective is in the anomalous position of arguing, "That is your point of view; ours is different. Let us put the difference to the test."

It seems to me far better to embrace the logic of the second alternative by holding that the "ought" in our conclusion is strongly pancultural, universally binding. We then can accept the values of freedom and tolerance: "Let these people be, for they have the right to freedom from interference, so long as they do not threaten others. This right, incidentally, is shared by all persons irrespective of culture."

I would argue for a parallel to this valuing of human freedom in the valuing of wilderness. Lands left to nature are home to many species of wild plants and animals. Some are threatened or endangered, others are the kith and kin of wild species existing in abundance outside wilderness lands. Whatever their status, these species enjoy a fundamental moral claim to freedom.

Whatever final form a pancultural view of wilderness takes as it is negotiated among nonauthoritarian cultures, I believe freedom will be its ultimate value. For nonhuman life, freedom can only be based on the spontaneity of wildness. Though wild species are genetically programmed (as is all of life), their repertoire of actions and reactions is astonishingly large, as the burgeoning science of ethology is only beginning to disclose. The case for extending the good of freedom to nonhumans derives from the suggestion that wild animals can remain what they are only if guaranteed noninterference by human beings. Under this view, noninterference may become a human duty well before wild animals are threatened with extinction. Preserving freedom for wild animals becomes a duty whenever a species cannot pursue its typical behavior in the nonwilderness environments where intensive human management is the norm—for example, in agricultural, pastoral, or urban places.

To defend this view, however, is to pass from demonstrating the possibility of a pancultural view of wilderness to arguing for its adoption, given that it is possible. I shall content myself with the observation that contemporary relativistic epistemologies do not mandate the conclusion that a pancultural view of wilderness is incoherent. In debating the relevance of the wilderness ideal to their own historical, cultural, and geographical circumstances, human beings from throughout the world show their understanding of the concept. Accordingly, we should conclude that a pancultural view is both possible and emerging.

Notes

The author is indebted to Valerie Chamberlain and Douglas Lind for helpful comments on earlier drafts of this paper.

1. Wallace Stegner, "The Wilderness Idea," in *Wilderness: America's Living Heritage,* ed. David Brower (San Francisco: Sierra Club Books, 1961), 97.

2. Harold K. Eidsvik, cited in John C. Hendee, George H. Stanley, and Robert C. Lucas, *Wilderness Management* (Golden, Colo.: North American Press, 1990), 90.

3. "Postmodernism has disclosed the contingency of all human speech-writing-thought (since it is predicated upon language), and perhaps in so doing has confirmed the insight of those sages who wrote the Kena Upanishad." Max Oelschlaeger, *The Idea of Wilderness* (New Haven: Yale University Press, 1991), p. 353. If I interpret him correctly, Oelschlaeger intimates that religion rather than philosophy must guide those who carry the wilderness banner into the future. My argument seeks to make philosophy at least a coequal in the quest.

4. Derrida resolutely refuses to state his views as doctrine. Instead, he tries to disclose by style and mannerisms (e.g., a fondness for puns, ambiguity, reversals of meaning, and neologisms)—how any interpretative enterprise is embedded in interpretation itself. His relativism is based on a deep suspicion of logic and reason. A good primer in English on his views is *A Derrida Reader: Between the Blinds,* ed. Peggy Kamuf (New York: Columbia University Press, 1991). Those wishing a more systematic exposition of Derrida-inspired relativism should consult

Stanley Fish, *Is There a Text in This Class? The Authority of Interpretative Communities* (Cambridge, Mass.: Harvard University Press, 1980).

5. I cite Quine's views specifically. For the others, see Nelson Goodman, *Ways of Worldmaking* (Indianapolis: Hackett, 1978), and Richard Rorty, *Philosophy and the Mirror of Nature* (Princeton, N.J.: Princeton University Press, 1978).

6. Willard Van Orman Quine, *Word and Object* (New York: John Wiley and Sons, 1960), 51-57, 73-79.

7. Luther Standing Bear, *Land of the Spotted Eagle* (Boston: Houghton Mifflin, 1933), 38.

8. Hendee et al. underscore the initial strangeness of wilderness preserves to indigenous peoples by suggesting that the case is analogous to proposing an "urban preserve" to a city-dwelling populace. See *Wilderness Management*, 92.

5 . Reminiscing about a Sleepy Lake: Borderland Views of Women, Place, and the Wild

Lois Ann Lorentzen

Our poetry cannot afford to be enamored with the moon when
we have bomb craters in our own land dropped by one of the
biggest and most powerful nations of the planet. Our poetry can
take the luxury of reminiscing about a sleepy lake or volcano, but
we cannot consent to being complicitous to the dangerous game
that proliferates sleepy consciousness.[1]

Women in Tehri, India, gathered on World Environment Day in 1979 with
empty water pots to protest water scarcity and the failure of water supply
projects. They said to the district collector, "We have come to tell you that
nature is the primary source of water, and we are the providers for our families. Unless the mountains are clothed with forests, the springs will not
come alive."[2] They recognized that, as Vandana Shiva writes, "the right to
food today is inextricably linked to the right of nature to conserve her ability
to produce food sustainably."[3] Culturally and socially constructed roles that
involve women in the day-to-day meeting of basic needs have often led to
environmental activism as women have made the link between environmental degradation and increased difficulties in providing sustenance.

This essay describes the participation of women in popular environmental movements to explore the claims made by thinkers such as Shiva that
Third World women, "whose minds have not yet been dispossessed or colonized," are in a privileged position to help create new intellectual ecological paradigms.[4] Third World women engaged in grassroots endeavors such
as the Chipko Andolan movement in India, ecofeminist movements in El
Salvador, and the Green Belt movement of Kenya may possess knowledge
and power not immediately accessible to the privileged because they know
directly what it means to be victims of progress. Third World women are
given culturally constructed roles that involve them in basic sustenance,
and they bear a disproportionate brunt of environmental impacts, so it

71

makes sense that many grassroots environmentalist movements are women-centered.

I will argue here that lessons learned from environmental movements in less affluent countries in general, and from ecofeminist movements in particular, suggest a desire for a sense of place that is neither fully "nature" nor fully "wilderness," but an ambiguous borderland. These movements often emerge out of concern over sustainability. Land reform is often critical to the struggle.[5] The challenge to ownership patterns is not to set aside land as wilderness for its own sake, but to promote a healthy ecosystem that includes humans. Many environmental activists from less affluent countries would argue that wilderness cannot be "saved" without land reform, redistribution of wealth, attention to gender, and protection of sustainable lifeways.

Although I speak here primarily about the environmental movements of the South, I also explore my own experience as a wilderness guide and professor of environmental ethics. In this context, I have observed that women's environmental activism in the North often stems from the "authentic" experience of self experienced in the wild and an existential need for nature to serve as a release from the stresses of civilization.

The Price of Development

The destructiveness of mainstream development models that link agriculture to capitalist-industrial development has been extensively documented and analyzed. Movements such as the Green Revolution removed agriculture from the context of a nature/survival economy and placed it within the market economy. New technologies were created not to protect soil and its fertility, but to fuel highly visible, state-financed production for profit. In this process fertility, pest control, and the growing and storing of seeds became no longer the work of peasants (especially peasant women), but the domain of agribusiness. Whereas traditionally peasants integrated care of animals and the forest with farming, the Green Revolution disrupted the links between animal husbandry and agriculture.

Mainstream development models did not lead to the promised food self-sufficiency (for instance, Africa now struggles to feed its people though it was food self-sufficient as late as 1970). This occurred for several reasons. Crop mixtures often were replaced by monocultures creating surpluses at the macro level while the poorer sections of rural society were actually worse off. Seed varieties were replaced, resulting in (for example) a radical decrease in the thousands of rice varieties found before 1980 in India. The crisis of desertification in Africa and other parts of the globe can be traced to characteristics of the Green Revolution, particularly the large monoculture and uniform cropping patterns, high nutrient uptake and low nutrient return to the soil by hybrid crops, high water demand and low water conservation of the new crops, waterlogged deserts in areas of large water projects, and water depletion where rivers were dammed or diverted.

Gail Omvedt lists the characteristics of capitalist agriculture as (1) a high use of external inputs brought over long distances by state and transnational agencies, in contrast to the locally produced, biologically based inputs upon which traditional systems rely; (2) a high degree of production for consumption elsewhere; (3) the predominance of monoculture over diversified cropping; (4) uneven development characterized by islands of seeming prosperity contrasted with "backward," drought-prone regions; (5) the encroachment on local decision-making power and local control of production by national and international forces; (6) the intensification of local inequalities; and (7) exploitation in the Marxist sense as well as damage to ecology and human health.[6]

In short, mainstream development has been a disaster both in terms of stated goals of increasing local self-sufficiency, and in terms of soil degradation, salinization, and dessication. These defects of market-driven agriculture have been analyzed in depth. What has received less attention is their disproportionate impact on peasant women. Resistance on the part of Third World peasant women has increased in the past decade as a result of failed development models.

Women and Land

> The virgin coffee
> dances in the shed.
> They strip it,
> rape it,
> out on the patios
> as it gives itself up to the sun.
> As the dark warehouses brighten,
> you can see malaria,
> blood,
> illiteracy,
> tuberculosis,
> misery

ripening in the golden coffee . . .
Country, etcetera,
country wound
child,
sob,
obsession.[7]

Salvadoran poet Claribel Alegria links violence to the earth with violence
to women. Many activist women from less affluent countries claim that environmental degradation affects women directly because their conceptual
link with nature makes them part of the dominated and exploited ecosystem.[8] An essentialist argument is not being made here. Nature, wilderness,
and woman's relation to nature/culture are seen as socially, politically, and
culturally constructed. When Salvadoran activist Rosario Acosta likens the
"sterilization of the land" to the "sterilization of woman," she is claiming
that under the social construction of export-oriented growth and capitalism
as experienced in El Salvador, women become conceptually linked with nature and thus are available as objects for appropriation and exploitation.[9]
Although the ecological problem affects all, women are especially hard hit
given their often unrecognized and uncompensated yet culturally pre-
scribed domestic work. Women are found on the forefront of environmentalist movements in many regions of the world because they are directly involved in family sustenance.

In rural India, women's traditional work was producing sustenance. With
the commodification of food production under the Green Revolution, the
role of women shifted from farm producer to subsidiary worker. Although
women's work in providing sustenance increased, the environmental damage meant longer work hours to provide basic needs and the economic value
of this work was discounted because it was linked to sustenance rather than
profit.

The impact on women of India's White Revolution mirrors that of the
Green Revolution. Through the White Revolution the cow was no longer
seen in a holistic fashion, providing fertilizer and energy as well as milk. Animal husbandry became rather the production of milk for the dairy industry. Women traditionally had been experts in animal husbandry as well as in
food processing, making curds, butter, ghee, and buttermilk. Two-thirds of
the power requirements of villages was met by work animals, and manuring
was critical in building up pest resistance in plants and in replenishing the
soil. In the reduction of the cow to a milk machine, basic rural needs were
neglected. Currently, 70 percent of the milk in India is manufactured into

products such as cheese, butter, and chocolates that are consumed by 2 percent of the population.[10] It is no surprise that the Chipko women of the hill areas of Garhwal resist the sale of milk.

It makes sense to focus on women when considering alternatives to misguided development models for several reasons. In Central America, women generally are food providers and producers, roles that depend on a healthy environment. With 80 percent of the natural vegetation of El Salvador eliminated and 77 percent of the soil severely affected by erosion and low soil fertility, it is increasingly difficult for peasant women to procure fuel, food, and fodder.[11] Without adequate forestation more hours are spent in fuel gathering and food production. The traditional diet of beans and corn, rice and tortillas is also undermined by the intensive marketing of processed foods such as bread and cola. Women bear the brunt of child care and care for the sick and elderly, so they are especially affected by contaminated water. As Alma Carballo of the Centro Salvadoreno de Tecnologia Apropriada (CESTA) notes, when polluted waters increase the incidence of cholera, "who deals with this at the hospital?"[12]

Gita Sen suggests that development, social and economic crises, and the subordination of women are closely linked.[13] According to her, we can most clearly witness the nature of the links in the chain of oppression, whether of rural peasant women or of nature, by examining the experience of the most oppressed. Women's relative access to economic resources, land, and employment has worsened as their burden of work has increased and their health, nutrition, and educational status have declined. Thus, because rural peasant women are the most affected, and presumably the most oppressed, they provide a clear vantage point as we assess the impact of development models and revision the human/nature relationship.

"My people today are weeping and the earth laments"[14]

What do initiatives spearheaded by women environmentalists suggest for the topic at hand—the idea of the wild, the human/nature relationship, and the notion of wilderness as a culturally relative term? The understanding of environmentalism is clearly not a direct import from more affluent nations. Many Third World environmentalists agree with Ramachandra Guha's claim that "the anthropocentrism/biocentrism distinction is of little use in understanding the dynamics of environmental degradation" and that "despite its claims to universality, deep ecology is firmly rooted in American environmental and cultural history and is inappropriate when applied to the

Third World."[15] This is seen most clearly in debates over the civilization/ nature relationship and language concerning wilderness and wilderness preservation.

Mainstream environmental groups from more affluent countries often argue for a notion of wilderness as a region free of human habitation and worthy of preservation in this state for its own sake. This notion is foreign to Chipko Andolan, the Green Belt movement of Kenya, and ecofeminists in El Salvador. Although a few elites and government officials might favor a notion of nature or wilderness as that locked away from human interference, the concept is largely scorned by grassroots environmentalist movements in less affluent countries, seen as the desire by the wealthy for "pretty places."[16] Whereas deep ecologists from more affluent nations may use the language of wilderness to evoke the place where one may experience the "wild" in an authentic manner, critics from the South view wilderness as a culturally and historically produced term. The concept of nature must be contextualized, and the notion of nature/wilderness as understood by mainstream environmentalists from the North does not reflect the reality of survival experienced by rural peasant female environmentalists.

The first time I encountered resistance to biocentric thinking in El Salvador, I was surprised. Having spent years as a mountain guide in the Sierra Nevada range of California, I was a staunch wilderness preservationist. Wilderness, or "pure" nature as I thought of it, was where I was most fully myself, where I was truly alive, where *I* was wild. The stresses of the city were left behind. As China Galland writes of river running, "I was hell bent to run that river. I felt that I was going under in the aftermath of a divorce. Being civilized became an impossibility. The only fit place for me was the wilderness—the wilder the better."[17] I agreed with Galland and I would return to Los Angeles from the mountains with my head out the car window screaming—hating the civilization, the stress, the relentless urban environment.

High on a mountain pass, gazing at endless mountain ranges, drinking from a high-altitude stream, sleeping by a glacial lake, I felt wild and also connected to all around me—trees, rivers, flowers, and towering granite. As Galland says, "Going into the wilderness invokes the wildness within us all. This may be the deepest value of such an experience, the recognition of our kinship with the natural world."[18]

I later discovered Susan Griffin and an entire genre of literature that corresponded to the exuberance, wildness, at-homeness, and "natural" feeling I had in the wilderness. Griffin successfully merged for me identification with the earth and experience of the wild:

> I know I am made from this earth, as my mother's hands were made from this earth, as her dreams came from this earth and all that I know, I know in this earth, the body of the bird, this pen, this paper, these hands, this tongue speaking, all that I know speaks to me through this earth and I long to tell you, you who are earth too, and listen as we speak to each other of what we know: the light is in us.[19]

I was one of those northerners desperate that my sacred places be kept intact without human interference. I needed my nature untouched.

What I didn't realize then was that nature is indeed a social, political, and cultural construction. As Vandana Shiva notes:

> [T]he "virgin" view of nature necessarily goes hand in hand with the "whore" view of what is not virgin nature. However, nature is neither "virgin" nor "whore," and ecology is not just conservation. Production happens in nature, in the home, in our daily lives, and is not limited by the artificial production and creation boundaries of patriarchal economics and science. The separation between production and reproduction, between innovation and regeneration, has been institutionalized to deny women and nature a productive role in the economic calculus.[20]

I had been experiencing the artificial boundaries of which Shiva writes when traveling back and forth between nature the "whore" (Los Angeles) and nature the "virgin" (my beloved Sierras).

What the critics of the South were suggesting to me was that possibly both women and nature needed to be reconceptualized. As Ann Usher writes of displacement of peasant communities in Thailand, "The intellectual separation of forests into 'conservation' and 'economic' zones simply does not work . . . political dispossession is not only unjust, it is also ecologically unviable."[21] The conceptual spheres into which I had divided nature, preserved and degraded, fit my experience as a First World wilderness guide. What my friends from the South proposed was that in their contexts, healthy ecosystems were those that accommodated humans.

Resistance to defining nature as "preserve" or wilderness occurs for several reasons. Humans (especially women) and the environment must "work or travel together" says Isabel Vasquez, director of the National Coordinating Committee of Salvadoran Women (CONAMUS).[22] Thus, CONAMUS projects include planting fruit trees in areas with adequate water and trees that provide firewood and berries for soil retention in more arid areas.

Of central concern to grassroots environmental movements is that the poor will die with a damaged environment. The central issue for peasant women in environmental movements is literally survival. Peasant women often are the agricultural producers, and rural women bear most of the burden for reproductive tasks, housework, care of the sick and elderly, and child care. Since rural women depend on the land for family survival and thus directly depend on their immediate environment for the daily necessities of life, their environmentalist battles revolve around equitable access, support of sustainable agriculture, the development of food processing and storage methods that ease the demands on women's labor, species diversification, the conservation of fertile soils, appropriate use of water and fuel—in short, battles that link survival of the land with human survival. Concern for the land is linked with survival for humans: nature is not a space that is separate but rather is home.

Forests as home, for example, provide fuel, food, and fodder. The link between forests and food is clear to the Chipko women of the hills of India who sing, "Give me an oak forest and I will give you pots full of milk and baskets full of grain."[23] When Chipko members hug trees to protest the commercial exploitation of forests by outside contractors, they know that forestry, animal husbandry, and agriculture must remain linked in practice for their survival. When Kenya's National Council of Women enlisted rural people as tree planters to fight desertification and soil erosion, it did so to protect local communities. These initiatives give little room for a notion of the wild as a place apart or even for very clear demarcations between nature and culture.

In fact, many ecofeminists and environmentalists articulate a strengthening of nature's capacity as essential for embracing uniquely regional cultures. Increased concern for the health of the land is intimately linked to enriching cultural identity. Rosario Acosta claims that "recuperating the land is a process of recuperating the culture and the ideals that we need to take care of nature."[24] Chipko activists view their actions as defending cultural values, in which specific areas are often viewed as sacred. As Taylor and colleagues note, "Western observers of non-Western environmental movements often fail to appreciate the site-specific nature of the spirituality of many groups."[25] Thus, the work of non-Western environmentalists seems to blur the lines between nature and culture, or at least bring the notions closer together by claiming that strengthening the one is part and parcel of strengthening the other. At the very least, the nature/culture relationship is not viewed as oppositional.

It also makes sense, given the survival orientation and cultural stance noted here, that such environmental work involves struggling directly on a social, political, and economic level to change structures that oppress the poor as well as the land. Thus, grassroots, female-led environmentalist groups maintain close ties to other popular movements, unlike many environmentalist groups from more affluent Western nations.[26]

I noted earlier that Gita Sen suggests that in exploring the oppression of rural peasant women and of nature, we should examine the experience of the most oppressed. In India, Central America, and parts of Africa, the "lowest of the low" is typically a woman, and her practices are worth noting. The work demanded of a rural peasant woman by a patriarchal culture influenced by Western-style development locates her labor somewhere between nature and culture. Although an essentialist position (which claims that women are more "natural" or closer to nature), is not appropriate, the caretaking labor demanded of women in this particular cultural construction places them in the space that holds together the family and the environment. Because they mediate between the needs of family and the environment, rural peasant women embody a blurring or dissolving of a strict nature/culture dichotomy. They avoid reductionist accounts of the relationship between nature and culture, and reject both the claim, that nature should be preserved free from human intervention and the opposite claim, that nature exists for human consumption. The activities of rural peasant women engaged in environmental activism yield a knowledge that rejects both forms of this dualistic construction.

The space that is sought, then, is one that is neither fully culture nor fully nature. For peasant women environmentalists, living in an ecological community is defined in relation to an environment that can produce safe food, water, fuels, and health care. The voices of such women, who religiously, philosophically, and concretely through their activism link concerns of gender and the land, speak to us of sustainable ways of living on the planet and with each other.

Notes

1. Anonymous Salvadoran woman poet, in *Ixok Amar Go: Central American Women's Poetry for Peace*, ed. and trans. Zoë Angelsey (Penobscot, Maine: Granite Press, 1987), 149.

2. Vandana Shiva, *Staying Alive: Women, Ecology and Development* (London: Zed, 1988), 211.

3. Ibid., 178.

4. Ibid., 46.

5. See Bron Taylor et al., "Grassroots Resistance: The Emergence of Popular Environmental Movements in Less Affluent Countries," in *Environmental Politics in the International Arena:*

Movements, Parties, Organizations, and Policy, ed. Sheldon Kamieniecki (Albany: SUNY Press, 1993), 69-90.

6. Gail Omvedt, "Green Earth, Women's Power, Human Liberation: Women in Peasant Movements in India," in *Close to Home: Women Reconnect Ecology, Health and Development Worldwide,* ed. Vandana Shiva (Philadelphia: New Society Publishers, 1994), 100-102.

7. Claribel Alegria, in Angelsey, *Ixok Amar Go,* 156.

8. Mercedes Canas, "In Us Grows Life: An Ecofeminist Point of View," *Palabra de Mujer* 3 (1991): 6-10.

9. Rosario Acosta, Executive Committee of the Union of Salvadoran Workers. Interview with the author. San Salvador, El Salvador, 10 June, 1993.

10. Shiva, *Staying Alive,* 172.

11. From *El Salvador: Una Crisis Ecologia,* Centro Salvadoreno de Tecnologia Apropriada (CESTA) (San Salvador, El Salvador, 1993).

12. Alma Carballo, Centro Salvadoreno de Tecnologia Apropriada (CESTA). Interview with the author. San Salvador, El Salvador, 23 June, 1993.

13. Gita Sen, *Development, Crisis and Alternative Visions: Third World Women's Perspectives* (Oslo: Verbum, 1985).

14. Eva Margarita Ortiz Platero, "The Bird's Song," in Angelsey, *Ixok Amar Go,* 199.

15. Ramachandra Guha, "Radical Environmentalism and Wilderness Preservation: A Third World Critique," *Philosophy & Public Affairs* 11, (1989): 71.

16. Acosta, interview.

17. China Galland, *Women in the Wilderness* (San Francisco: Harper & Row, 1980), 10.

18. Ibid., 5.

19. Susan Griffin, *Woman and Nature: The Roaring inside Her* (San Francisco: Harper & Row, 1978), 227.

20. Vandana Shiva, ed. *Close to Home: Women Reconnect Ecology, Health and Development Worldwide* (Philadelphia: New Society Publishers, 1994), 5.

21. Ann Danaiya Usher, "After the Forest: AIDS as Ecological Collapse in Thailand," in Shiva, *Close to Home,* 37.

22. Isabel Vasquez, General Director of National Coordinating Committee of Salvadoran Women (CONAMUS). Interview with the author. San Salvador, El Salvador, 7 June, 1993.

23. Shiva, *Staying Alive,* 97.

24. Acosta, interview.

25. Taylor et al., "Grass Roots Resistance," 74.

26. I realize this may be an overgeneralization and largely ignores the work of groups influenced by social ecology.

6 . Confessions of an Eco-Colonialist: Responsible Knowing among the Inuit

Douglas J. Buege

Anthropologist and social philosopher Hugh Brody begins his book *The People's Land* with an anecdote concerning a planeload of white men, *Qallunaat* in the language Inuktitut, landing at an Arctic settlement in eastern Canada. Air traffic is a rare enough occurrence for word of these visitors to circulate rapidly throughout the Inuit community. Brody spent a half hour with these nine white males, who were looking for information concerning Arctic fishing conditions. In this brief time, he tells:

> [T]hey quizzed me about how the Eskimos fished for Arctic char near the set-tlement. I reminisced about a few fishing trips but my knowledge of the sub-ject was scanty. I repeatedly suggested that they talk to some of the local hunters who were at home because of the bad weather. A meeting with the best informants could easily be arranged, and I assured them that these [Inuit] hunters would be very interested to hear what they [the whites] had to say about char fisheries in other places. No, they said, their time was too short. But, I pointed out, their time here had been extended by the weather. . . . In the High Arctic, summer evenings last all night, no one feels bound to go to bed at any particular time and they could talk as long as they liked. No, they said, they had been invited to visit the home of one of the whites in the settlement; that night they would not have time for any prolonged discussion.[1]

Indeed, the white fishing enthusiasts spent the night at the other white man's house, apparently favoring cocktails and white companionship to the infor-mation they claimed to desire, information that would have been best sup-plied by the Inuit fisherpeople. The Inuit experts were slighted and knew that these visitors did not value their knowledge.

This story provides one example of how Inuit knowledge is ignored even when it is easily obtained and invaluable to the seekers of such knowledge, in this case, white men. Now I would like to consider another example that il-lustrates a much more insidious consequence of disregarding indigenous people's knowledge, a consequence that may not be as obvious as the earlier example.

Since 1955, various groups have protested the hunting of seals by dwellers of northern regions. During this time, the Inuit were moved by market forces to expand their hunting of ringed seals. The price of a ringed sealskin increased from less than $1 (Canadian) in 1960 to $15 in 1975 and $25 to $30 in the early 1980s, making seal hunting an increasingly lucrative activity.[2] Harvesting ringed seals came to be preferred to wage labor as a source of income for the Inuit, who were becoming increasingly reliant upon nontraditional goods and services.[3] Inuit hunters found that they could make a living hunting and trapping as long as they could use weapons and vehicles introduced by Euro-Americans.

The moral values of Euro-Americans who protested the killing of harp and hooded seals came to have an impact upon the Inuit, who were supposedly not a target of these protesters. In 1982, the European Economic Community (EEC) approved a ban on many sealskin goods, crippling the market for Inuit-harvested furs.[4] A representative of the International Fund for Animal Welfare defended the collapse of the sealskin market by arguing that his organization would not have a problem with Inuit hunting if they hunted in the same way they did "five hundred years ago."[5]

In effect, the Inuit people are caught in what feminist philosopher Marilyn Frye calls a double bind: they can maintain "traditional" ways exclusively and lose any economic power they have, or they can adopt tools such as rifles and relinquish any rights they have to harvest furbearing animals.[6] Both options are undesirable: the former reduces Euro-Americans' resistance in exploiting the resources of the Arctic, while the latter basically denies the Inuit their heritage and reduces them to the status of generic, colonized "others." Inuit are forced to choose between these two options by outsiders to their culture, people who see the need to impose their morality upon the Inuit.

Pressured by the popular animal rights movement, the EEC failed to consult the Inuit concerning the ban. As in the fishing example, Inuit knowledge was ignored, but this time the Inuit were the people most adversely affected. While the EEC is relatively stable in the face of economic adversity, the Inuit economy is less stable because it is greatly influenced by forces beyond Inuit control. It is only fair to include the Inuit in deliberation of actions that affect them.

Both examples illustrate some of the problems with intercultural communication of knowledge. The fifth World Wilderness Congress focused on environmental issues specifically concerning wilderness in circumpolar regions, so I've chosen to discuss some of the issues that involve the Inuit of

Canada's Eastern Arctic, a group that has inhabited Arctic regions for more than 4,000 years. When I discuss Inuit knowledge, I view such knowledge as political, cultural, and ecological.

While certain legal changes have improved the political status of Inuit peoples (e.g., the Inuit are represented in the Canadian Parliament by officials they elect, and the majority of people managing the Canadian Arctic are Inuit), these changes have not ended colonial action in the Canadian Arctic any more than civil rights legislation has ended racism in the United States. Indeed, a colonial mentality still pervades Euro-American society: we have stereotypes that inform our conceptions of the Arctic and the people who live there and we still lust after the rich resources hidden below the Arctic landscape. The reforms in Canada have merely forced us to adjust our colonial actions, not abandon them. This essay reflects my own struggles with colonialism as a privileged white Euro-American male and discusses some ways in which I have come to understand my own colonial actions since visiting the Arctic five years ago. I believe that some readers may recognize aspects of themselves in my discussion and may be able to recognize some of the ways in which they participate in the ongoing projects of colonialism.

"Primitive" Ideologies

Before non-Inuit people can begin to understand the experiences of any Arctic cultural groups, we must examine our perceptions of these people. We have preconceived notions of what life is like for dwellers of the Arctic. We also have misconceptions of what the Arctic is as a physical environment.[7]

In the 1920s, filmmaker Robert Joseph Flaherty made and released the documentary *Nanook of the North*. This silent film was a box-office smash in the United States, Canada, and Europe. Flaherty describes his intentions in making the movie: "I wanted to show the Inuit. And I wanted to show them, not from the civilized point of view, but as they saw themselves, as 'we the people.'"[8]

Despite Flaherty's intentions, people were drawn to the exotic culture of the Eskimos depicted in the film and romanticized a lifestyle so incredibly different than their day-to-day existence in Cleveland, Paris, or wherever. Visions of Nanook standing on the ice in caribou skins thrusting a spear mesmerized audiences and invited them to fantasize about life in the Arctic. This film presented Nanook and his kin not as political, intellectual beings, but as savages with subsistence lifestyles, emphasizing the severity of such existence. Stunned by the harsh environmental conditions, audiences could not avoid being awed by this severity. I contend that many of the perceptions of Nanook's culture this film fostered are still alive today, more than seventy years after Nanook himself starved to death.

Marianna Torgovnick discusses the perceptions we hold of peoples such as the Inuit in her book *Gone Primitive*. She explains how the idea of primitive cultures elicits many widely held preconceptions:

> Primitives are [said to be] like children. . . . Primitives are our untamed selves, our id forces—libidinous, irrational, violent, dangerous. Primitives are mystics, in tune with nature, part of its harmonies. Primitives are free. Primitives exist at the "lowest cultural levels"; we occupy the highest, in the metaphors of stratification and hierarchy commonly used [by some anthropologists].[9]

She continues by arguing that these conceptions of primitive people do not arise out of a knowledge of these people; instead, such ideologies are created by outsiders, Euro-Americans, in order to shape the Inuit and other groups into something we desire. Perhaps we are thrilled by the idea of savages living in a severe climate under the harshest of environmental, social, and economic conditions. If that is what we desire, that may be how we view the Inuit.

An interesting aspect of Euro-American ideologies of the Arctic is the fact that few people have traveled north of the Arctic Circle. Most Euro-Americans receive their information of the Arctic from secondary sources such as Flaherty's film and various written accounts. Such accounts are instrumental in creating the stereotypes associated with native peoples. Mary Louise Pratt discusses some of the more popular and successful ways in which written accounts can create a stereotyped "other":

> The people to be othered are homogenized into a collective "they," which is distilled even further into an iconic "he" (the standardized adult male specimen). This abstracted "he/they" is the subject of verbs in a timeless present tense, which characterizes anything "he" is or does not as a particular historical event but as an instance of a pregiven custom or trait.[10]

Whether one examines travel literature, ethnographic studies, or less academic adventure stories concerning the Arctic, Pratt's homogenized he/they is readily apparent. One consequence of accounts written in the manner Pratt describes is that Euro-Americans conceive of Arctic peoples as male-dominated and primitive. The "timeless present tense" creates the belief that Inuit people have been doing the same activities for centuries, their lives supposedly remaining unchanged.

Another way in which stereotypes of the Inuit are informed is through descriptions of the Arctic as a physical environment. Euro-Americans have vaguely developed ideas concerning the physical conditions of the far North, with ice and snow dominating our mental images. Indeed, ice and snow are key components of the landscape, as are rock, wind, clouds, and water. Yet more revealing are the adjectives and metaphors we use to describe this physical environment. (In another paper, I examine prevalent views of the Arctic by several writers.[11] Some of the most common adjectives used by these authors are "bleak," "lifeless," "desolate," and "featureless.") Once we accept the severity of the Arctic as a fact, we make many assumptions of the people who live there: these people are different from Euro-Americans, they are rugged and savage, they live ascetic lifestyles, bereft of comfort and security.

When we combine expectations of the people based on the environment they inhabit with other tools for creating stereotypes such as the literary devices discussed by Pratt, we develop some very potent stereotypes that prevent us from seeing Inuit people as they might want to be seen. We have many expectations linked with these stereotypes. This is because our conceptions of indigenous people are not merely descriptive; they also carry normative force.

When I visited Baffin Island, Northwest Territories, back in 1988, I was taken aback by the acned teenage Inuit wearing acid-washed blue jeans and leather bomber jackets, playing video games. I didn't want the Inuit to be like people back home in Wisconsin. After growing up with all the stereotypical images of the "Eskimos" (who rubbed noses to express affection toward one another), I expected these people to be rugged, dressed in the caribou skins that I knew so well. My vision of what it means to be Inuit involved my deciding what an Inuit *should* be. Such ideas, when expressed, tend to undermine the self-determination of the Inuit people. They also expose my prejudices.

One particularly potent ideology concerning indigenous peoples is the idea of the "ecologically noble savage."[12] We are led to believe that the Inuit are particularly ecologically responsible people, that they are in harmony with their natural environments. But such a viewpoint cannot stand up to scrutiny. The Inuit took to guns and ammunition, motorboats, snowmobiles and television very quickly. Many are willing to exploit the natural resources on their land. The North Slope Inupiat of Alaska have taken strong prodevelopment stances prompted by the possibility of gaining economically from the sale of oil from their land, and there are few reasons why Canadian Inuit should not do the same.[13] The conception of the ecologically noble savage is challenged by much evidence, yet seems to be maintained.[14]

Inuit people may be justified in capitalizing upon the wealth of resources from their lands. The economic means for survival outweigh the importance of living up to the Euro-American stereotype of "ecological nobility." Given the history of Euro-American relations with resource colonies such as the Arctic, it is highly likely that U.S. and Canadian companies will find ways to extract the mineral, oil, and natural gas of the Arctic regardless of Inuit interests. Thus, the Inuit sale of such resources is not a sign of ecological irresponsibility; instead, it should be seen as a sign of growing economic and political prudence.

The main problem with the ideology of the ecologically noble Inuit is that it leads us to expect a certain type of behavior from the Inuit, a behavior that they do not live up to. Our stereotype often proves damaging to their cultural integrity. To expect the Inuit to return to traditional hunting ways is to impose our values upon them, values that we obviously do not hold as strongly for our own societies. We cannot expect the Inuit to abandon their snowmobiles and motorboats, along with all the other luxuries of a market economy, in order to live up to our expectations.

A more significant problem that stems from viewing the Inuit as primitive is that such a perspective comprehends the Inuit as children, as people who need to be taken care of in this modern world, a world as foreign to them as it is familiar to Euro-Americans.[15] The result of this view is that we find a need to minister to their religious, political, educational, and economic needs. The Inuit then tend to rely upon the systems we set up to benefit them, creating a self-perpetuating cycle of dependence that seems to confirm our original ideas of the primitive as child. In effect, the ideology of the primitive serves as an apparently irrefutable rationalization for patriarchy.

"Responsible Knowing"

In an earlier paper, I developed a theory of knowledge based upon Lorraine Code's "responsible knowing."[16] I connected Code's theory with environmentalism to create a position I call *environmentally responsible knowing*. Responsible knowing focuses upon individuals as knowers who have a responsibility to obtain and use knowledge in activities in which they participate, and are accountable for that knowledge. There are five points to responsible knowing: knowers, emotions, understanding, normative realism, and community. It is not my goal to argue that the Inuit are environmentally responsible knowers; to do so would negate a project that I find most important, that of using ongoing dialogue to determine knowers' environmental responsibilities. I will have more to say about this later. But I will show how these five points relate to the Inuit living in their Arctic environment.

Knowers

Traditionally, theories of knowledge (epistemologies) have failed to emphasize the importance of individual knowers. The view of responsible knowing, however, places the knower central to the examination of knowledge claims. Knowers know in particular contexts, so time, place, culture, and environment all shape how knowers perceive the world. Inuit knowers are heavily influenced by their Arctic environments and their social structures. Their knowledge is also affected by colonial actions of Euro-Americans. The EEC, in its sealskin moratorium, failed to take contextual issues of the Inuit into account.

In environmentally responsible knowing, individuals become more environmentally responsible by developing relationships with their environments—not simple one-way relationships, but interactive ways of knowing. As Caroline Whitbeck argues, the self/other opposition commonly used to define relationships is inadequate.[17] Instead, I endorse Whitbeck's mutual realization, a view that holds that reciprocity is a fundamental part of relationships. Responsible knowing is demonstrated by information, often in the form of narrative, that illustrates a mutually beneficial relationship between knowers and their environments.

Emotions

Responsible knowing takes emotions as central to cognitive practice. Emotions are an essential component of oppressed people's knowledge of the world; they "enrich [their] knowledge of the nature of that oppression," according to Uma Narayan.[18] When Hugh Brody discusses the Qallunaat fishermen, he mentions how the Inuit felt slighted by the attitudes of the whites. This emotional response is significant because it shows that the Inuit were affected by the white men's impertinence even though they might not be able to offer a verbal account of what happened. Their emotional response may be quite complex: they might blame themselves for being less than accommodating, they may feel their way of life is less important, they may feel powerless to react, they may feel anger toward the Qallunaat that they do not want to express. All these emotional reactions are important in understanding the larger picture of Inuit perceptions of their world.

Understanding

Understanding is a complex way of knowing. It involves creating a holistic picture of many individual facts and processes as they relate to one another. Understanding one's natural environment involves knowing how activities affect that environment and how one can live in that environment sustainably. A well-developed knower/environment relationship is characteristic of an understanding of one's environment.

Narrative can illustrate one's understanding of one's environment. Minnie Aodla Freeman combines her understandings of her environment and her experience of colonialism:

> Inuit never went out into the ocean without testing their kayak first, Inuit never put up their igloo without examining the location, Inuit did not go into action without weighing the total situation first. The plane arrives, the government or industry officials step out and out comes a new situation. Often, even today, no letters, no phone calls, no information. Are *Qallunaat* always so unthinking, unfeeling and so rash?[19]

This narrative does more than simply point out the rudeness Aodla Freeman experiences with Qallunaat; it also exhibits an understanding of how Inuit and Qallunaat are, and how they relate to one another. Her words carry much more information because she invests these words with her personal perspective as a woman and an Inuit who has lived among Qallunaat for more than half of her life.

Normative Realism

The condition of normative realism simply means that we should work to understand the world as it is and not as we may want it to be. I have already brought up one example of where it is important to strive for normative realism: though we may wish to view the Inuit as primitive or as ecologically noble savages, neither perspective is empirically justified. At all times we must take account of the perspective from which we view the world and work to understand how the world really operates.

In order to survive and keep their cultural traditions, the Inuit must struggle for normative realism. They need to understand how to maintain their own traditions while coexisting with First World colonialism. In order to stop environmental projects they challenge, they must know how to argue their cases in the Canadian government and through the media. They must understand much about white culture just to preserve their own culture.

Allootook Ippellie demonstrates normative realism when he writes, "It would be far better and more appropriate to grant power to those whose lives and the environment they live and depend on is being threatened."[20]

Community

Environmentally responsible knowers do not operate in a vacuum; they share their knowledge with others. The community is an important structure for gathering knowledge and transmitting it to future generations and outside peoples. Interaction of individuals in the community promotes honesty and development of language and ethical concepts.

Language is a key device for maintaining community. It is interesting to notice how Inuktitut captures much information that assists in maintaining social ties. For example, place-names often offer up a depiction of the location. A place English-speakers call "Baker Lake" the Inuit call "Qamanituaq," which translates to "where the river widens into a lake."[21] Such descriptive naming promotes the sharing of travel itineraries, aiding Inuit in traveling vast stretches of land accurately. Indeed, the ability to travel in the Arctic depends upon learning from experienced elders how to navigate and to use place-names.

Colonialism threatens the traditional social structure of the Inuit, thus affecting their conveyance of traditional knowledge. Peter Inukpuk reports, "Many of our young people are losing the ability to live off the land, and in this future situation, we will all live on welfare, strangers in the land of our forefathers."[22] Education, speaking English, and the increased availability of affordable Western goods all contribute to the acculturation of younger people. Paradoxically, at the same time the young Inuit are learning Western ways that help them understand how to maintain Inuit autonomy, they risk losing touch with their own culture.

Inuit Knowledge

My criticism of the ideology of the ecologically noble savage does not mean I believe that Inuit are as environmentally destructive as typical Americans or Europeans. Certainly the Inuit people have gathered an incredible amount of information about their particular environments. An environmentally responsible knowledge of one's environment imparts an expertise in dealing with issues concerning that environment. In the case of the eastern Canadian Arctic, the Inuit people have the most expertise in dealing with their surroundings because they have lived there for a very long time and their knowledge of the land is more extensive than that of any other group. The Inuit certainly have developed very strong interactive relationships with their land.

In relation to my goal to show that the environmental situation of the Inuit is linked to political, social, and cultural views that are continually ignored, I'd like to offer two more narratives. This first was written by Inuit author Akeeko more than thirty years ago:

> Last summer, Eskimos were asked about moving to the DEW Line. Some did not want to go and some wanted to. I know what help they really need. They need money. Possibly some could get money by trapping foxes. But foxes are not plentiful. One has to go very far to get a fox. To do this, it is necessary to leave a wife and children behind without fuel to burn. It is not like it was long ago. Some Eskimos hunt well and some don't. I know myself now how hard it is to do what I used to do. It is also hard not understanding and yet sort of knowing the meaning in English. I have never been to school. Some Eskimos work very hard and some don't. Some have sense and some don't. Others help each other. This is the way it is in our country.[23]

This personal appeal illustrates much about the changes the Inuit have experienced and are yet experiencing. Connections to the land are now linked to the market economy imposed upon the Inuit, eroding the social ties of kinship groups and communities. Akeeko also mentions the loss of traditional hunting skills that he observes in himself and in others.

It may be easy for some readers to dismiss Akeeko's knowledge as an artifact of the early 1960s. We certainly would like to believe that conditions have improved for the Inuit since Akeeko's thoughts were recorded on paper. But seventy-three-year-old Simon Akpaleeapik echoes Akeeko's concerns today. Akpaleeapik, who lives in Pond Inlet, a small community toward the northern end of Baffin Island, remarks:

> We had nothing until I was 10 years old. We wore only caribou skins, but now we have these things [Euro-American goods]. . . . The only thing that's changed is the younger generation is taking all this for granted . . . they don't know anything. When I was 10 to 15 years old, I watched my father. We weren't taught, but we learned by watching to hunt, butcher, skin and use a harpoon.[24]

Apparently, conflicts still exist between elder Inuit, who retain traditions, and the younger people, who accept their acculturation uncritically. Akpaleeapik foresees the loss of much of the Inuit culture in which he grew up. What may make this loss even more tragic is that we are likely to miss the value of much of this cultural knowledge for Inuit and Euro-American alike only when it is gone forever. We have little understanding of the long-term effects of acculturation upon people such as the Inuit.

Are the Inuit environmentally responsible knowers? This is a difficult question to answer simply. I believe that in the past, before contact with Qallunaat, they certainly were. Their lifestyle was sustainable, intimately linked with the natural environment, yet not overtly disruptive of that environment.

Today conditions are quite different. Thanks to colonialism, Inuit are in a state of flux. Environmental responsibility is connected to all aspects of Inuit lifestyles. If the Inuit become a dependent culture that occupies government-owned lands, then I doubt they can be environmentally responsible people. In other words, the power of colonialism is pernicious enough to rob a people of their most important qualities, their responsibility and their relationships. Some Inuit may be able to maintain environmentally responsible relationships to the land and its occupants; others will lose these connections as they become acculturated.

On the other hand, if the Inuit can obtain a high degree of autonomy over their lands and their culture, then I believe they will have the basis for maintaining intimate relationships with their environment. The Nunavut agreement, which divides the Northwest Territories into two provinces, one being the 136,000 square mile Nunavut, is one step toward such autonomy.[25] Certainly more needs to be done.

But the ability for Inuit people to attain this state of autonomy depends upon how Euro-Americans react to the Inuit. If we cling to our stereotypes of the Arctic and the Inuit, then we are likely to continue our colonialistic ways, treating the Inuit like children who need direction from us. The stereotypes that have influenced my life have altered my perceptions of the Inuit in the past. Now I am working to change these false ideas and understand how such stereotypes are created and how they function. I hope that some of the directions I have offered here help other colonialists to understand themselves and motivate them to change their behavior. Our ignorance has great impact upon all colonized peoples.

Conclusion

Earlier in this essay I referred to ongoing dialogue as one project I would like this work to inspire; now I will conclude with some comments concerning dialogue. Inuit are no longer isolated from Euro-Americans; thus, Euro-Americans should not remain isolated from the Inuit and their concerns.

Both groups have important information to share. Euro-Americans cannot understand themselves without understanding how colonialism affects groups such as the Inuit. Dialogue between individuals and between cultures is essential for understanding one another. For the most part, this dialogue depends upon Euro-Americans respecting and listening to voices they have traditionally ignored. This dialogue can and should be entered by everyone concerned with the Inuit and the future of the Arctic.

Notes

I would like to give my fondest thanks to Dr. Helen Hoy, Yvonne Holl, Jason Demeny, David Rothenberg, David Buege, the staff of the Historical Society Library at the University of Wisconsin, and the people of Broughton Island and Pangnirtung, Northwest Territories, for making this work possible.

1. Hugh Brody, *The People's Land: Eskimos and Whites in the Eastern Arctic* (Harmondsworth: Penguin, 1975).

2. George Wenzel, "Marooned in a Blizzard of Contradictions: Inuit and the Anti-Sealing Movement," *Etudes/Inuit/Studies* 9 (1985): 75-91.

3. It is important to note that hunting, a traditional activity, is far preferential to wage labor. In part this is because traditional activities are more fulfilling and essential to cultural survival than are wage positions, which are commonly services rendered to non-Inuit people. Hunting is an activity of an *Inummariit*—in English, a "real Eskimo."

4. "Anti-Sealing Lobby Severely Hurting Inuit Hunters," *Native Perspective* 2:7 (1977): 8.

5. Daniel Morast, in Wenzel, "Marooned," 84. Wenzel offers a plethora of such statements from various proponents of animal welfare and environmentalism.

6. Marilyn Frye, "Oppression," in *The Politics of Reality* (Freedom, Calif.: Crossing Press, 1983).

7. See Brody, *People's Land,* for an in-depth discussion of the connection between images of the land and of the people. On page 81 he comments, "The great social and intellectual distance between Whites and Eskimos is emphasized in the minds of Whites by the harshness of the Arctic and the intimate closeness of Eskimo life with the land: the harsher the environment, the closer to nature must be the people who are able to inhabit it."

8. As quoted in Richard Griffith, *The World of Robert Flaherty* (Westport, Conn.: Greenwood, 1953).

9. Marianna Torgovnick, *Gone Primitive: Savage Intellects, Modern Lives* (Chicago: University of Chicago Press, 1990).

10. Mary Louise Pratt, "Scratches on the Face of the Country; or, What Mr. Barrow Saw in the Land of the Bushmen," *Critical Inquiry* 12 (1985): 119-43.

11. Douglas J. Buege, "Frozen in Place: Euro-American Ideologies and the Inuit," paper delivered at the Interdisciplinary Research Network on the Environment and Society's conference, "Perspectives on the Environment 2: Research and Action," University of Sheffield, September 1993.

12. For an extended discussion of the ideas discussed here see Kent Redford, "The Ecologically Noble Savage," *Cultural Survival Quarterly* 15 (1991): 46-48.

13. Richard Eathorne, "Fade to Black: The Untold Story in the ANWR Battle," *Buzzworm* 3 (1991): 34-35.

14. Euro-American perpetuation of the ideology of the ecologically noble Inuit is possible because we deny the economic and cultural problems the Inuit face. By denying these problems, we fail to recognize that Inuit, in order to maintain their ways of life, must focus on cer-

tain issues that are more essential for their immediate and long-term survival. It is not surprising that in difficult times, such as today when colonial actions bring great cultural change to the Inuit, economic issues prevail over environmental issues.

15. The arrogance we have of ourselves as experts on surviving in the modern world is exemplified by former U.S. president George Bush's infatuation with the creation of a "new world order," quite obviously an order that denies the self-determination of almost all peoples on the planet, including many citizens of First World countries as well as all Third and Fourth World peoples.

16. Douglas J. Buege, "Epistemic Responsibility to the Natural: Toward a Feminist Epistemology for Environmental Ethics," *APA Newsletter on Feminism and Philosophy* 91 (1992): 73-78.

17. Caroline Whitbeck, "A Different Reality: Feminist Ontology," in *Women, Knowledge, and Reality,* ed. Ann Garry and Marilyn Pearsall (Boston: Unwin Hyman, 1989), 51-76.

18. Uma Narayan, "Working Together across Differences: Some Considerations of Emotions and Political Practice," *Hypatia* 3 (1988): 31-47.

19. Minnie Aodla Freeman, "Living in Two Hells," *Inuit Today* 8 (1980): 32-35.

20. Allootook Ippellie, "Editorial," *Inuit Today* 9 (1981): 4-5.

21. David F. Pelly, "How Inuit Find Their Way in the Trackless Arctic," *Canadian Geographic* 3 (1991): 58-64.

22. Quoted in Eric Alden Smith, "Inuit of the Canadian Eastern Arctic," *Cultural Survival Quarterly* 8 (1984): 32-37.

23. Akeeko, "Akeeko Is Writing," in *Paper Stays Put,* ed. Robin Gedalof (Edmonton: Hurtig, 1980), 14-20.

24. Quoted in Storer H. Rowley, "Arctic Traditions Become a Matter of Survival," *Chicago Tribune,* 20 February 1994, sect. 1, 23.

25. For a good presentation of Nunavut see Jim Bell, "Nunavut: The Quiet Revolution," *Arctic Circle* (1992): 12-21.

III . THE ART OF THE WILD

7 . Out of the Map, into the Territory: The Earthly Topology of Time

David Abram

> This very time that is space, this very space that is time, which I will have rediscovered by my analysis of the visible and the flesh . . .
>
> *Maurice Merleau-Ponty*

These reflections are motivated by a hunch: the suspicious intuition that time and space are not really distinct dimensions for any animal, for any animate body, human or otherwise. Yet I am puzzled. Why do my friends and fellow philosophers assume that space and time *are* distinct dimensions of the world, or even distinct dimensions of their experience? Where is there any element within the sensorial world that suggests or promotes this division between the temporal and the spatial? Everywhere I cast my focus I find space timing or time spacing. How, then, did these powers come to be separated in so perplexing a manner?

I wish to write of two ways whereby what we name "time" may be felt to metamorphose into what we call "space." In the course of these metamorphoses, "space" itself is transformed into something very different from what we now assume.

The Opening-Outward of Time

I would like to begin by pondering, in a somewhat cursory manner, the prospects for an ecological, postmodern interpretation of death. Such an interpretation has not, to my knowledge, been formally articulated, and I will not do so here. Yet it is anticipated in the view of death expressed in the songs and practices of indigenous, oral cultures. An ecological interpretation of dying would be one that recognizes the living not as a determinate mechanism but as a magic webwork of reciprocities and cyclical relations between diverse symbiotic entities—diverse cells, tissues, organs—all engaged in an ecstatic, improvisational dance. The body, acknowledged as an interdependent and shifting lattice of relationships, is a living ecosystem, a microcosm of the encompassing order, or equilibrium, in which it is embedded.

Death, then, might be interpreted as that moment when the central equilibria of the body begin to break down, the moment at which this field of recursive relations no longer maintains itself as something relatively separate or autonomous, but instead begins to open outward into the larger field of relations—into the soils, the waters, and the air—opening outward, that is, into *space*, the very realm that envelops me as I write this.

Of course, death, in indigenous cultures, has its mystery and its terror, yet this is not because it is anticipated as the impending cessation of existence. Death, for oral peoples, is not felt as the approach of nonbeing or nothingness. Rather the awesomeness of death is that of an astonishing metamorphosis, the imminence of one's reincorporation into the encompassing cosmos. Those who have died are not elsewhere—they are present in the very space that enfolds us, in the ripples that move through a flock of birds turning in the sky, in the sounds of a stream, or the wind sighing in the grasses. Death, in such a culture, is not the closure of time, but a gateway into environing space. Indeed, it is these same spatial depths from whence we each emerge at the moment of birth. The dead and the not-yet-born, the past and the future, are somehow present in this very space that sensorially envelops us.

Yet this is not to do away with *time* entirely, for then we would be left with a space that exists all at once, with no room for events, with no sound or reverberation, no heart that could get so far as to beat, no breath that could be drawn. Or, conversely, we could describe a space without time as a zone where *all* events happen at once, all blossoming in the same moment, all beats of the heart together, all breaths both inhaled and exhaled instantaneously. All of which amounts to the same thing—that is, a space that is conceived *without depth*, without a near and a far, without a place from whence it could encounter itself, without hollows wherein Being withdraws and from which it emerges. Such a depthless space has nothing to do with our actual experience, nothing to do with the life of our eyes or our ears—it has no room for the hand that is writing the word "space" or for the tongue that says it. This non-sense is what we conceive when we think space as something separable from time—whether we imagine this space as an utterly empty void without phenomena, or as an entirely filled plenum or pure extension without gaps, holes, or folds.

But these are mere games, tricks we play with ourselves when we pretend that we have no body, these words not really written on a page. We can play these games precisely because we are *in* depth. (Depth loves to play games.

Think of how an airplane roaring in the distance turns out to be a tiny fly buzzing near your nose.) I can pretend not to be a body precisely because this body dwells in depth and thus has access to various forms of concealment and invisibility: the hiddenness of its interior; the concealment of its shoulder blades, which it can touch but cannot see; even the subtle invisibility of the very medium that it ceaselessly breathes in and out—as though I might drift out of myself with my exhalations, the body floating out of itself with the breeze. And of course it does! Thus I meet aspects of myself out in the world: staring back at myself from the surface of a window late at night. Sometimes I forget myself back in the car and have to go back to find myself; on certain mornings I've even flushed myself down the toilet. But although these are ways of absenting myself they never escape the flesh entirely—I can slip out with my breath into the air and thus play at invisibility, but the air is itself a tactile and olfactory being, and so while the eyes may be fooled into thinking that I have gone, the nose and the skin will not be fooled. Unless, of course, I make that grander escape from the density of this body into verbal ideas and concepts. Then I am present neither to the eyes nor to the skin, but the *ears* will not be fooled, for they can hear me sounding in my skull, and even if I flee on wings of speech into space, they can hear me still, in the speech of the crows overhead, or the rattling of branches, or the crunch of a stone falling into snow, and so they will know that I am still placed in the world—out wandering for a time, but not far, *not all gone.*

Sometimes it is the wild earth itself who draws us out, coaxing us into its depths, making us run, leap, and cry in delight at this vastness, ducking into holes or swinging on branches or just lying back on the grass and staring at stars; looking up through the swirling lens of earth's sky, gazing out across the solar depths, beyond Saturn, beyond Neptune, floating out into the very Body of the galaxy, grazing the stars with our eyes, our fingers brushing the grass, our mouths drinking the night air. Earth does this to us sometimes, or the wind does it, turning our whole body inside out so that we are outside ourselves without having to leave, finding ourselves, by virtue of being in *this* place, suddenly connected to *all* places. It is this that makes my words wonder if there's an eternity to be had not by getting *beyond* this world but simply by being *in* it, not by escaping the body but by *becoming* the body.

Such an experience is neither of time nor of space, but of earthly *place*— of this breathing world as a locus of mysteries in which we participate.

Placing Time

When I returned to North America from a year among tribal peoples in Indonesia and Nepal, I found myself perplexed and confused by many aspects of my own culture. Assumptions that I had previously taken for granted, or that I had, since childhood, accepted as obvious and unshakable truths, now made little sense to me. The belief, for instance, in an autonomous "past" and "future." Where *were* these invisible realms, which had so much power over the lives of my family and friends? Everybody that I knew seemed to be expending a great deal of effort thinking about and trying to hold on to the past—obsessively photographing and videotaping events—and continually fretting about the future—ceaselessly sending out insurance premiums for their homes, for their cars, even for their own bodies. As a result of all these past and future concerns, everyone appeared (to me in my raw and newly returned state) to be strangely unaware of happenings unfolding all around them *in the present*. They seemed oblivious to the myriad nonhuman lives that surrounded them, to the minute gestures of insects and plants, the speech of birds, the flux of sounds and smells. My family, my old friends all seemed so oblivious, now, to the sensuous *presence* of the world. The present, for them, seemed little more than a point, an infinitesimal *now* separating "the past" from "the future." And indeed, the more I entered into dialogue with my family and friends, the more readily I, too, felt my consciousness cut off, as though by a sheet of reflective glass, from the life of the land.

There is a useful exercise that I devised back then to keep myself from falling completely into the civilized oblivion of linear time. You are welcome to try it the next time you are outdoors. I locate myself in a relatively open space—a low hill is particularly good, or a wide field. I relax a bit, take a few breaths, gaze around. Then I close my eyes and let myself begin to feel the whole bulk of my past—the whole mass of events leading up to this very moment. And I call into awareness, as well, my whole future—all those projects and possibilities that lie waiting to be realized. I imagine this past and this future as two vast balloons of Time, separated from each other like the bulbs of an hourglass, yet linked together at the single moment where I stand pondering them. And then, very slowly, I allow both of these immense bulbs of Time to begin leaking their substance into this minute moment between them, into the present. Slowly, imperceptibly at first, the present moment begins to grow. Nourished by the leakage from the past and the future, the present moment swells in proportion as those other dimensions shrink.

Soon it is very large; and the past and future have dwindled down to mere knots on the edge of this huge expanse. At this point I let the past and the future dissolve entirely. And I open my eyes . . .

I find myself standing in the midst of an eternity, a wild and inexhaustible present. The whole world rests within itself—the trees at the field's edge, the hum of crickets in the grass, cirrocumulus clouds rippling like waves across the sky, from horizon to horizon. In the distance I notice the curving dirt road and my rusty car parked at its edge—these, too, seem to have their place in this open moment of vision, this eternal present. And smells—the air is rich with faint whiffs from the forest, the heather, the soil underfoot—so many messages mingling between different elements in the encircling land. The jagged snag of a single withered oak tree standing alone in the field does not, in this eternity, seem really dead. It is surrounded by an admiring clump of bushes, and a large boulder reposes at the edge of those bushes, discoursing with the old tree about shadows and sunlight.

Stepping closer, I can see that the crumbling bark around the oak's trunk is crossed by two lines of ants, one moving up the trunk and the other heading down into the soil. From this closer vantage I can see, too, that the shadows on the boulder are not really shadows at all, but patches of lichen spreading outward from various points on the rock's surface, in diverse textures and hues—dull blacks and crinkly grays and powdery, deep reds—as though through them the rock was silently expressing its inner moods. I scratch my leg. Strangely, the vividness of this world does not dissipate. I stomp on the ground, spin around, even stand on my head. But the open present does not disperse. Several jet-black crows race out of the woods, chasing each other in swoops and sudden dives; one of them lands on the crumbling snag. "Kahhr! Kahr! Kahr!" Now it glides down to the ground just in front of me,—"Kahr!"—and stands there looking at me, sideways, through a purple eye. The lids blink swiftly, like shutters. It hops around me and the big beak opens. "Kawhhr!" I try to reply, "Cawr!" and the bird steps forward. Crow does not hop, I see, but walks, clumsily, on this ground. I can see the tiny feathers covering the nostrils on its beak as the breeze picks it up off the ground, feel myself swoop through the swirling breeze toward the forest edge.

Things are different in this world without the past and the future, my body quivering in this space like an animal. I know well that, in some time out of this time, I must return to my house and my books. But here, too, is home. For my body is at home, in this open present, with its mind. And this is no mere illusion, no hallucination, this eternity—there is something too persistent, too stable, too unshakable about this experience for it to be merely a mirage.

The unshakable solidity of this experience is curious, indeed. It seems to have something to do with the remarkable affinity between this temporal notion that we term the "present" and the spatial landscape in which we are embedded. When I allow the past and the future to dissolve, imaginatively, into the immediacy of the present moment, then the present itself expands to become an enveloping field of *presence*. And this presence, vibrant and alive, spontaneously assumes the precise shape and contour of the enveloping sensory landscape, as though this were its native shape! It is this remarkable *fit* between temporal concept (the present) and spatial percept (the enveloping *presence* of the land) that accounts, I believe, for the relatively stable and solid nature of this experience, and that prompts me to wonder whether time and space are really as distinct as I was taught to believe. There is no aspect of this realm that is strictly temporal, for it is composed of spatial things that have density and weight, and is spatially extended around me on all sides, from the near trees to the distant clouds. And yet there is no aspect, either, that is *strictly* spatial or static—for every perceivable being, from the stones to the breeze to my car in the distance, seems to vibrate with life and sensation. In this open present, I am unable to isolate space from time, or vice versa. I am immersed in the world.

What, then, has become of the past and the future? I found my way into this living expanse by dissolving past and future into the sensorial present that envelops me; did I thereby do away with them entirely? I think not. I simply did away with these dimensions as they are conventionally conceived—as autonomous realms existing apart from the sensuous present. By letting past and future dissolve into the present moment, I have opened the way for their gradual rediscovery—no longer as autonomous, mental realms, but now as aspects of the corporeal present, of this capacious terrain that bodily enfolds me. And so now I crouch in the midst of this eternity, my naked toes hugging the soil and my eyes drinking the distances, trying to discern where, in this living landscape, the past and the future might reside.

The French philosopher Maurice Merleau-Ponty, in one of the notes found on his desk after his untimely death in 1961, addressed the same conundrum:

> In what sense the visible landscape under my eyes is not exterior to . . . other moments of time and the past, but has them really *behind itself* in simultaneity, inside itself, and not it and they side by side "in" time.[1]

And so we are faced with this puzzle: Where, within the visible landscape, can we locate the past and the future? Where is their place in the sensuous world?

Of course we may say that we perceive the past all around us, in great trees grown from seeds that germinated long ago, in the eroded banks of a meandering stream or the widening cracks in an old road. And, too, that we are peering into the future wherever we look—watching a storm cloud emerge from the horizon, or a spiderweb slowly taking shape before our eyes—because, in a sense, all that we perceive is already pregnant with the future. But how, then, can we *distinguish* these two temporal realms? We certainly have a sense that the past and the future are not the same; nevertheless, they are strangely commingled within all that we perceive. How, then, do they distinguish themselves perpetually? If we say that the past is where all that we see *comes from* and the future is where it is all *going*, we simply beg the question, naming two allegedly obvious domains that we remain unable to locate within the perceivable landscape—as though past and future are, indeed, pure intuitions of the mind, existing in some incorporeal dimension outside of the sensible world. This, presumably, is what prompts many scientists and philosophers to assert that other animals have no real awareness of time—no sense of a past or a future—since they lack any intellect that could apprehend this hidden dimension.

As an animal, I remain suspicious of all these dodges, all these ways whereby my species lays claim to a source of Truth that lies outside of the bodily world wherein plants, stones, and streams have their being—outside of this earthly terrain that we share with the other animals. And yet, as a philosopher, I feel pressed to account for these mysteries, these "times" that are somehow *not present,* these other "whens." And so now let us bring the human animal and the philosopher in ourselves together, and try to locate the past and the future within the sensory landscape. And let us proceed with caution, for by now we can see that there is a great deal at stake in our investigation: the imputed distinction between space and time; the distance of the literate intellect from the bodily senses; the alleged contrast between humans and all other animals.

First, we shall take some methodological guidance from Merleau-Ponty who, more than thirty years ago, was already struggling to give voice to "this very time that is space, this very space that is time."[2] In his last work, Merleau-Ponty describes the relation between the perceptual world and the world of our supposedly incorporeal ideas and thoughts: "It is by borrowing from the world's structure that the universe of truth and of thought is constructed for us."[3] These words assert the primacy of the bodily world relative to the universe of ideas; they suggest that the structures of our apparently incorporeal ideas are lifted, as it were, from the structures of the perceptual world. If we read Merleau-Ponty's words carefully, and accept their guidance, we discern that what we are hunting for here, in our deepening quest, are specific structures in the perceptual landscape that have lent their particular character, or shape, to these two persistent ideas, the past and the future. We are searching, that is, for a structural correspondence—an isomorphism, or match—between the *conceptual* structure of the past and the future and the *perceptual* structure of the surrounding sensory world.

If we have taken a kind of method from Merleau-Ponty, it is to Martin Heidegger that we should turn for a careful structural description of the past and the future. Throughout his life, from his first to his final writings, Heidegger gave special attention to the phenomenon of Time, and it was he, more than any other thinker, who developed a phenomenology of Time's dimensions.

We should give special attention to two works that seem to bracket his lifelong investigations, his early book *Being and Time,* and his late essay "Time and Being."[4] In the middle of the essay we find Heidegger asking the very question we ourselves have posed: "Where is time? *Is* time at all and does it have a place?"[5] He then goes on to distinguish that time into which he is inquiring from the common idea of time as a linear sequence of "nows":

> Obviously, time is not nothing. Accordingly, we maintain caution and say: there is time. We become still more cautious, and look carefully at that which shows itself as time, by looking ahead to Being in the sense of presence, the present. However, the present in the sense of presence differs so vastly from the present in the sense of the now. . . . the present as presence and everything which belongs to such a present would have to be called real time, even though there is nothing immediately about it of time as time is usually represented in the sense of a succession of a calculable sequence of nows.[6]

Heidegger's philosophical move, here, to disclose behind the present considered as "now" a deeper sense of the present as presence, approximates our own experiential move to expand the punctiform "now" by dissolving the past and the future as conventionally experienced, thereby locating ourselves in a vast and open present—which we, too, have called "the present as presence." According to Heidegger, it is only from within this experience of *the present as presence* that "real time" or primordial time (which later in the essay he will call "time-space") can make itself evident. In our case the present has determined itself as presence only by taking on the precise contours of the visible landscape that enfolds us. We are now free to look around us, in this vast terrain, for the place of *the past* and of *the future.*

And Heidegger offers us a helpful clue. In *Being and Time,* he writes of past, present, and future as the three "ecstasies" of time. He seems to mean that the past, the present, and the future all draw us outside of ourselves. Time is ecstatic in that it opens us outward. Toward what? The three ecstasies of time, according to Heidegger, "are not simply raptures in which one gets carried away. Rather, there belongs to each ecstasy a 'whither' to which one is carried." Each of time's ecstasies carries us, Heidegger says, toward a particular horizon.[7]

As soon as we pay heed to this curious description, we notice an obvious correspondence between the conceptual structure of time, as described by Heidegger, and the perceptual structure of the enveloping landscape. The horizon itself! Heidegger, in *Being and Time,* uses the term "horizon" as a structural metaphor, a way of expressing the enigmatic nature of time. Just as the action of time seems to ensure that the perceivable present is always open, always already unfolding beyond itself, so the distant horizon seems to hold open the perceivable landscape, binding it always to what lies beyond it.[8]

The visible horizon, that is, is a kind of gateway or threshold, joining the presence of the surrounding terrain to what *exceeds* this open present— to what is hidden *beyond* the horizon. The horizon carries the promise of something more, something *other*. Here we have made our first discovery: the way that other places, places not explicitly present within the perceivable landscape, are nevertheless joined to the present landscape by the visible horizon. And so let us ask whether it is possible that the realms we are looking for, the perceptual place of *the past* and of *the future,* are precisely beyond-the-horizon.

Certainly this is a useful first step. For clearly, neither the past nor the future is entirely out in the open of the perceivable present, *and yet they seem everywhere implied.* Since the horizon effectively implicates all that lies beyond the horizon within the present landscape that it bounds, it seems plausible to suppose that both the past and the future reside beyond the horizon.

Yet this leaves me somewhat confused, for I am unable, then, to account for the *difference* between the past and the future. The horizon of the perceivable landscape is provided, I know, by the relation of my body to the vast and spherical body of the earth. This is not merely something that I have read, or learned in school. It has become evident and true for me in the course of many journeys across the land, watching the horizon continually recede as I move toward it, watching it disgorge unexpected vistas that expand and envelop me even as the horizon itself withdraws, maintaining its distance. And yet if I glance behind me as I journey, I see that this enigmatic edge is also *following* me, keeping its distance behind me as well as in front, gradually swallowing those terrains that I walk, drive, or pedal away from. May I then conclude that *the future* is beyond that part of the horizon toward which I am facing, while *the past* is beyond that part of the horizon that lies behind me? Then I would need only to turn around in order for my past to become my future, and vice versa. Hmmm . . . This does not seem quite right. If I journey toward the horizon—toward any *part* of that horizon—I will indeed disclose new places and things that were previously in my future, beyond the horizon. Certainly I can attempt the reverse, as when I journey back toward that distant town where I used to dwell. But in this I am never quite successful. For that town, when I arrive, is no longer as it was. The old schoolhouse now stands half-collapsed in a field overgrown with wildflowers and thistles; the marsh where I used to await, each spring, the arrival of herons has vanished beneath a huge shopping mall. The land has changed. I cannot, it seems, journey toward the past in the same way that I can journey toward the future. For the past does not *remain* past beyond the horizon; it does not wait for me there like the future.

It is this strange asymmetry of past and future in relation to the present that Heidegger describes in "Time and Being." Whereas in *Being and Time*, Heidegger wrote of the centrifugal, ecstatic character of time—of time as that which draws us outside of ourselves, opening us to what is other—in this later essay he stresses the centripetal, inward-extending nature of time, describing time as a mystery that continually approaches us from beyond, extending and offering the gift of presence while nevertheless withdrawing behind the event of this offering. Such descriptions may sound strange— even uncanny—to our ears, and yet we should listen to them closely. For as Heidegger's thought matured, he increasingly sought to loosen awareness from the bondage of outworn assumptions precisely by wielding common words in highly unusual ways, shaking terms free from their conventional usages. Thus *past* and *future* are here articulated as hidden powers that approach us, offering and opening the present while nevertheless remaining withdrawn, concealed from the very present that they make possible. In Heidegger's description, both the past and the future remain hidden from the open presence that they mutually bring about.[9] And yet the way that the future conceals itself in its offering is quite *different* from the manner in which the past is concealed in its giving. Specifically, the future, or what is to come, *withholds* its presence, while the past, or what has been, *refuses* its presence.[10] The future *witholds;* the past *refuses.* In its most complete description of the vicissitudes of time, Heidegger puts the matter thus (and let us listen to his words from within the open presence of the land around us):

What has been which, by *refusing* the present, lets that become present which is no longer present; and the coming toward us of what is to come which, by *withholding* the present, lets that be present which is not yet present—both [make] manifest the manner of an extending opening up which gives all presencing into the open.[11]

We must acknowledge the weirdness of Heidegger's language here (he is trying to avoid the use of nouns, of nominative forms that would freeze the temporal flux). It is a strangeness, however, that enables his words to approach, and to open us onto, the silent structuration of this mystery we call time. Pondering these words in the midst of the open landscape—from within the green and windswept presence of the present—leads us now to ask: Where can we perceive this withholding and this refusal of which Heidegger speaks? Where can we glimpse this refusal and this withholding that open and make possible the sensuous presence of the world around us?

We have already noticed the magic by which the horizon encloses and yet holds open the visible landscape: precisely by concealing or—better—withholding what lies beyond it. Thus the horizon may indeed be felt as a withholding. But it is hardly a refusal. Its lips of earth and sky may touch one another, but they are never sealed, and we know that if we journey toward that horizon it will gradually disclose to us what it now withholds.

Where, then, can we locate the refusal to which Heidegger alludes? Do we perceive such a refusal anywhere around us? More important, how do we even know what we are looking for? Here, again, we take a clue from Heidegger. In "Time and Being," he writes of *absence* as a modality of presence. More specifically, he writes of the past and of the future as absences that *by their very absence* concern us, and so make themselves felt within the present.[12] This description aids us a great deal. Now at least we can say what we are searching for in our attempt to locate, or place, the past and the future. We are hunting for modes of absence that, by their very way of being absent, make themselves felt within the sensuous presence of the open landscape. Or in Merleau-Ponty's terminology, we could say we are searching for certain invisible aspects of the visible environment, unseen regions whose very hiddenness somehow structures the open visibility of the world around us. *The beyond-the-horizon is just such an absent or unseen realm.* And so we must now ask: Is there *another* unseen aspect, another absent region whose very concealment is somehow necessary to the open presence of the landscape?

Of course there are those facets that I cannot see of the things or bodies that surround me—the sides of the trees that are facing away from me, or the other side of that lichen-covered rock. Yet these concealments are all analogous, in a sense, to what lies hidden beyond the horizon. The other side of that rock, for instance, is *withheld* from my gaze, but it is not *refused,* for I can disclose it by walking over there, just as I can disclose what lies beyond the horizon by a longer journey.

What of my own body? Well, most of my body is present to my awareness, and visible to my gaze—I can see my limbs, my torso, and even my nose—although my back, of course, is hidden beyond the horizon of my shoulders. The back of my body is withheld from my gaze, and yet I know that it is present—visible even to the crows perched behind me in the trees, as I know that the fields and forests hidden beyond the horizon are yet visible and present to those who dwell there.

Yet while pondering the unseen aspect of my body, I soon become aware of another unseen region: that of the whole *inside* of my body. The inside of my body is not, of course, *entirely* absent; but it is hidden or concealed from visibility in a manner very different from the concealment of my back, or of what lies beyond the horizon. It is an instance, I suddenly realize, of a vast mode of absence or invisibility entirely proper to the present landscape—an absence I had almost entirely forgotten. It is the absence of what is *under* the ground.

Like the beyond-the-horizon, the absence of the under-the-ground is an absence so familiar, so necessary to the open presence of the world around us, that we take it entirely for granted, and so it has been very difficult for me to bring it into awareness. But once I have done so, the recognition of this hidden region begins to clarify and balance the enigmatic power of that other unseen region beyond the horizon.

For these would seem to be the two primary dimensions from whence things enter the open presence of the landscape, and into which they depart. Sensible phenomena are continually emerging out of, and continually receding into, these two very different realms of concealment or invisibility—the realm beyond the horizon and the realm under the ground. Although they balance one another, these two double-vectors—receding into or emerging from the beyond-the-horizon; sinking into or arising out of the under-the-ground—are by no means symmetrical. The one is a passage outward toward, or inward from, a vast openness. The other is a descent into, or a sprouting up from, *a packed density.* While the horizon *withholds* the visibility of what lies beyond it, the ground is much more resolute in its concealment of what lies beneath it. It is this resoluteness, this *refusal* of access to what lies beneath the ground, that enables the ground to solidly support all those phenomena that move or dwell upon its surface. Thus although the absence of the beyond-the-horizon and the absence of the under-the-ground reciprocate one another, they contrast markedly in their relation to the sensuous terrain. We may describe this reciprocity and this contrast thus: The beyond-the-horizon, by withholding its presence, holds open the perceived landscape, while the under-the-ground, by refusing its presence, supports the perceived landscape. The reciprocity and asymmetry between these two realms bear an uncanny resemblance to the reciprocity and contrast between the future (what is to come) and the past (what has been) in Heideggers's description: the one *witholding* presence, the other *refusing* presence, both of them thus making possible the open presence of the present. Dare we suspect that these two notions describe one and the same phenomenon? I believe that we can, for the isomorphism is complete.

The Invisibility of the Air

By reading Merleau-Ponty and Heidegger together, and by setting their words in relation to our own experience, we have discerned that the past and the future—these curious dimensions—may be just as much *spatial* as they are *temporal*. Indeed, we have begun to *place* these dimensions, to discern their location within the sensuous world that enfolds us. The conceptual abstraction that we commonly term the future would seem to be born from our bodily awareness of what is hidden beyond the horizon—of what exceeds, and thus holds open, the living present. What we commonly term "the past" would seem to be rooted in our carnal sense of what is hidden under the ground—of what resists, and thus supports, the living present. As ground and horizon, these dimensions are no more temporal than they are spatial, no more mental than they are bodily and sensorial.

We can now discern just how close Merleau-Ponty, himself, was to this discovery by reading his aforementioned note of November 1960 in the light of our disclosures:

> In what sense the visible landscape under my eyes is not exterior to, and bound systematically to . . . other moments of time and the past, but has them really *behind itself* in simultaneity, *inside itself* and not it and they side by side "in" time.[13]

For we can now understand this *behind* and this *inside* in a remarkably precise manner. The visible landscape has the other moments of time "behind itself," precisely in that the future waits beyond the horizon—as well as *behind* every entity that I see, as the unseen "other side" of the many visibles that surround me. And the visible landscape has the other moments of time "inside itself," precisely in that the past preserves itself under the ground—as well as *inside* every entity that I perceive. The sensorial landscape, in other words, not only opens onto that distant future waiting beyond the horizon but also onto a near future, onto an immanent field of possibilities waiting behind each tree, each stone, behind each leaf from whence a spider may at any moment come crawling into my awareness. And this living terrain is supported not only by that more settled or sedimented past under the ground, but by an immanent past carried inside each tree, inside each blade of grass, within even the tissues and cells of my own body (and the bodies of the other animals and insects around me).

It is thus that ecologists and environmental scientists may study the recent past of a particular place by "coring" several of the standing trees, in order to count their interior rings and to interpret the varying width of those rings (an extrawide layer, thirteen rings in from the cambium, suggests a season of abundant rain thirteen years into the depth of the past, whereas an extrathin layer tells of a year without rainfall). The deeper past may be pondered by digging a "soil pit" to expose the sedimented layers of the soil, and to interpret the composition and structure of those layers (a layer of charcoal, for instance, bespeaks a forest fire at that depth of the past). Meanwhile, archaeologists, paleontologists, and geologists dig still deeper beneath the ground of the present in order to unearth traces of ancient epochs and eons.

That which has been and that which is to come are not elsewhere—they are not autonomous dimensions independent of the encompassing present in which we dwell; they are, rather, the very depths of this living place—the hidden depth of its distances and the concealed depth on which we stand.

The past is the ground of the living present; the future is the horizon of the living present. Yet here we must acknowledge a strange ambiguity. The beyond-the-horizon is that realm where the sun goes when it leaves us, and the realm from which it emerges at dawn; it is where the moon goes and returns from. But we could just as well say "the sun sinks into the under-the-ground, the moon emerges from the under-the-ground." For when we attend closely to our direct, sensory experiences of the rising and the setting, we see that the moon's journey beyond the horizon is also experienced as a movement down into the ground, and indeed that the sun's rise each morning is as much an emergence from under the ground as is the emergence of a groundhog at the end of winter! Hence, for example, these words by the Kiowa author, N. Scott Momaday:

> "Where does the sun live?" . . . to the Indian child who asks the question, the parent replies, "The sun lives in the earth." The sun-watcher among the Rio Grande Pueblos, whose sacred task it is to observe, each day, the very point of the sun's emergence on the skyline, knows in the depths of his being that the sun is alive and that it is indivisible with the earth, and he refers to the farthest eastern mesa as the "sun's house" . . . Should someone say to the sun, "Where are you going?" the sun would surely answer, "I am going home," and it is understood that home is the earth. All things are alive in this profound unity in which are all elements, all animals, all things . . . my father remembered that, as a boy, he had watched with wonder and something like fear the old man Koi-khan-hole, "Dragonfly," stand in the first light, his arms outstretched and his painted face fixed on the east, "praying the sun out of the ground."[14]

Phenomenologically, it is as though the luminous orb of the sun journeys into the ground each evening, moving all night through the density underfoot, to emerge, at dawn, at the opposite side of the visible world. For some oral cultures, it is precisely during this journey through the ground that the sun impregnates the earth with its fiery life, giving rise to the myriad living things—human and nonhuman—that blossom forth on earth's surface.

We begin to glimpse, here, the secret identity, for oral peoples, of those topological regions that we have come to call "the past" and "the future"— the curious manner in which these very different modes of absence transmute into each other, blur into one another like moods. It is thus that many indigenous cultures have but a single term to designate the distant past and the distant future. Among the Inuit of Baffin Island, for example, the term *uvatiarru* may be translated both as "long ago" and "in the future."[15] The cyclical metamorphosis of the distant past into the distant future, or of that-which-has-been into that-which-is-to-come, would seem to take place continually, in the depths far below the visible present, in that place where the unseen lands beyond the horizon seem to fold into the invisible density beneath our feet.

As well as right here, in the wildness of the open present. Here, too, this living realm where we stand is born of the continual convergence and metamorphosis of past and future into a kind of eternity. Martin Heidegger did not, in fact, write of only *two* temporal ecstasies, but of *three*, including that of the present. In *Being and Time*, Heidegger asserts that the present, itself, has *its own* ecstasy, its own proper transcendence, its own "'whither' to which one is carried away."[16] And in "Time and Being," Heidegger writes that "even in the present itself, there always plays a kind of approach and bringing about, that is, a kind of presencing."[17] The implication is that the present can be concealed not just within the past or the future, but within the very depths of itself. As though, paradoxically, there is a modality of absence entirely native to the present, another mode of absence or invisibility not beneath or beyond the present but at the very heart of the living present, from whence the present, itself, comes to presence: "In the present, too, presencing is given."[18]

Is there, then, yet another mode of absence or invisibility entirely endemic to the open presence of the surrounding earth? I have already noticed, here within the perceivable present, the hidden nature of what lies behind the tree trunks and stones that surround me, which corresponds to the unseen character of those lands entirely beyond the horizon of the perceivable present, from whence numerous entities enter the visible terrain and into which various phenomena withdraw, recede, and finally vanish from view. I have acknowledged, as well, the concealed character of what rests inside the trunks of these trees, within the stones and the hills, which corresponds, ultimately, to the concealedness under the ground, from whence beings sprout and unfurl, and into which they also crumble, decompose, and are submerged. Is there some *other* obvious style of absence, in the very thickness of the present, that is unique to itself, and not a mere modification of the under-the-ground or the beyond-the-horizon? Some mode of concealment that is, paradoxically, already out in the open, from whence the living present itself comes to presence?

Perhaps I am pushing my method too far, here, in trying to place not only the withholding of presence by the future and the refusal of presence by the past, but also this concealment of presence *from within the present itself.* For now, more than ever, I feel confused—unable to grasp, or to conceive of, what it is that I am searching for. Even as I gaze out across the wooded hills, my mind seems muddled by these questions, by ideas and associations that keep me from directly sensing and responding to the animate earth around me. I try to relax, and so begin to breathe more deeply, enjoying the coolness of the breeze as it floods in at my nostrils, feeling my chest and abdomen slowly expand and contract. My thinking begins to ease, the internal chatter gradually taking on the rhythm of the inbreath and the outbreath, the words themselves beginning to dissolve, flowing out with each exhalation to merge with the silent breathing of the land. The interior monologue dissipates, slowly, into the rustle of pine needles and the stately gait of the clouds.

A butterfly glides by, golden wings navigating delicate air currents with a few momentary flutters before they settle on a scarlet flower. My sensing body now awake to the world, I gradually become conscious of a third mode of invisibility, of an unseen dimension in which I am so thoroughly and deeply immersed that even now I can hardly bring it to full awareness.

It is the invisibility of the air.

Of course the air, in one sense, is the most pervasive presence I can name, enveloping, embracing, and caressing me both inside and out, moving in ripples along my skin, flowing between my fingers, swirling around my arms and thighs, rolling in eddies along the roof of my mouth, slipping ceaselessly through throat and esophagus to fill the lungs, to feed my blood, my heart, my self. I cannot act, cannot speak, cannot think a single thought without the participation of this fluid element. I am immersed in its depths as surely as fish are immersed in the sea.

Yet the air, on the other hand, is the most outrageous absence known to this body. For it is utterly invisible. I know very well that there is something there—I can feel it moving against my face and I can taste it and smell it, can even hear it as it swirls within my ears and along the bark of trees, but still . . . I cannot see it. I can see the steady movement it induces in the shapeshifting clouds, the way it bends the branches of the cottonwoods and sends ripples along the surface of a stream. Yet I am unable to see the air itself.

Unlike the hidden character of what lies beyond the horizon, and unlike the unseen nature of what resides under the ground, the air is invisible *in principle*. That which today lies beyond the horizon can at least partly be disclosed by journeying into that future, as that which waits under the ground can be unearthed by excavations into that past. But the air can never be opened for our eyes, never made manifest. It remains, in this sense, the great mysterious.

And this mystery is what enables life to live. It unites our breathing bodies not only with the under-the-ground (with the rich microbial life of the soil, with fossil and mineral deposits deep in the bedrock), and not only with the beyond-the-horizon (with distant forests and oceans), but also with the interior life of all that we perceive in the open field of the wild present—the grasses and leaves, the ravens, the buzzing insects and the drifting clouds. What the trees are breathing out, we animals are breathing in; what we breathe out, the plants are breathing in. The air, we might say, is the soul of the visible landscape, the secret realm from whence all beings draw their nourishment. As the very mystery of the living present, it is the place of Time's central ecstasy (of that most intimate absence from whence the present presences), and thus a key to the forgotten presence of the earth.

When the animate earth is forgotten, the powers of the under-the-ground and the beyond-the-horizon are displaced. Abstract potencies without a home, in one epoch they are set in opposition to each other as "Hell" (the underworld) and "Heaven" (the promised land beyond the horizon—El Dorado or Shangri-La—projected into the sky). Then, in our own era, they are reified into pure concepts, "the past" and "the future," fictions of our contemporary language with no meaning for the living body, existing nowhere. Yet the condition for this final diminution of earth's generative mysteries into disembodied mental concepts was provided by the loss of the invisible richness of the present, by the forgetting of the air. For to forget the air was to forget the place of *psyche* in the body's world, and so to set the stage for the notion of "mind" as an exclusively human attribute. The word "psyche"—like the word "spirit"—comes from a term for the breath and the wind.

In this essay we have shown that there is at least one way to reconcile time and space perceptually. In doing so we have transformed space from a conceptual abstraction into this sensuous *place* in which we are corporeally immersed, this world structured by both a ground and a horizon, this earth. We now have a way of visualizing time without representing it as an arrow, or as a time line with a series of "now" points marked on it. The shape or contour of time is precisely the contour of this living field that envelops us. This time cannot be diagrammed from outside. It can only be entered and explored.

....

Notes

Borrowing Korzybski's metaphor, we have stepped out of the map, and into the territory.

1. Maurice Merleau-Ponty, *The Visible and the Universe*, trans. Alphonso Lingis (Chicago: Northwestern University Press, 1966), 267.

2. Ibid., 259.

3. Ibid., 13.

4. Heidegger's early book, *Being and Time*, trans. MacQuarrie and Robinson (New York: Harper and Row, 1962), analyzed the different dimensions of time solely as these enter into and structure human existence, while his much later essay, "Time and Being," in Heidegger, *On Time and Being*, trans. Joan Stambaugh (New York: Harper and Row, 1972), strives to elucidate the phenomena of time with regard to existence, or Being, itself. Since I am here concerned with the relation *between* these two foci—that is, between humans and the more-than-human field of existence—I will draw from both of these texts in what follows.

5. Heidegger, "Time and Being," 11.

6. Ibid., 11-12.

7. Heidegger, *Being and Time*, 416.

8. In truth the idea of "time" is a thoroughly horizon-laden thought for Heidegger; in *Being and Time* he can hardly mention the phenomenon of "time" in any capacity without linking it to the horizon metaphor. Thus, when explicating the genesis of our ordinary conception of time as a linear sequence, Heidegger translates Aristotle's definition of time in the following manner: "For this is time: that which is counted in the movement which we encounter within the horizon of the earlier and later" (*Being and Time*, 473). And indeed, the entire book ends with the question: "Does *time* itself manifest as the horizon of *Being?*" (*Being and Time*, 488).

9. Ibid., 13.

10. Ibid., 16-17.

11. Ibid., (emphasis added).

12. Ibid., 13, 17.

13. Ibid., 267 (the second emphasis is added).

14. N. Scott Momaday, "Personal Reflections," in Calvin Martin, ed., *The American Indian and the Problem of History* (New York: Oxford University Press, 1987), 156–61.

15. John James Houston, "Songs in Stone: Animals in Inuit Sculpture," *Orion Nature Quarterly* 4 no. 4 (Autumn 1985): 8.

16. Heidegger, *Being and Time*, 416.

17. Heidegger, "Time and Being," 15.

18. Ibid., 13.

By Hans Gerhard Sørensen.

8 . Silent Wolves: The Howl of the Implicit

Irene Klaver

Wilderness: you don't go there to find something, you go there to disappear.

Wallace Stegner

Silence . . . [is] a sort of resting place where they [words, objects] will finally disappear, silence which is no longer anything.

Georges Bataille

Wild is the smoke rising, turning in slow movements, tracing invisible currents of air with its gray white elegance before finally disappearing. Wild are the tracks of the mountain lion in the snow and the scratches on my arms and legs, painful results of trying to follow the animal through the thickets of its trajectory in fields and bushes. Traces. Implicit presences, referring to more than we can say and see: to be wild is to stand out *and* to disappear.

The deer standing out against the tree line has vanished as suddenly as it came to the fore, and it is perhaps impossible, and in any case irrelevant, to determine if the deer has leaped into the forest or if the surroundings have taken it up. What constitutes the wild and the silent is this very play between appearance and disappearance, the slipping in and out of the limits of presence. Untamed and not named, the wild and the silent escape the frames of our knowledge. Miles of fungal filaments in a handful of old-growth forest soil. The feeling of frost in the air, the cold touching my skin, the rich fragrance after a summer rain.

Writing is always responding. Responding to words, scratches, thoughts, events, the rain, the trees. Taking up what is implicit. Implicit in texts, words, things, gestures, scents. Sometimes by means of detailed analyses of other texts, other thinkers, political situations, specific predicaments; sometimes through rough sketches of what lingers in the air. Always showing the thickness, the different layers, of what is here. Yet we never get more than a fragment: "Life is the time the words need to enter the book, the time man has to exhaust words and embrace silence."[1]

Sitting outside. Jacob-dog-lying at my feet. Wind hiding in the trees. All quiet. Monday-cat-sleeping on my lap. My hands resting on her fur. "Fur," I whisper. Puff of air. That is all there is to it. Fur. Air. But words have to be found for the book. To break the silence that speaks for itself, without voice, *unbestimmt*, undetermined.

I

Besides being an area of land untrammeled by humanity, wilderness is a story. A relatively new story, historically idiosyncratic to the West, being told from the time civilization got a firmer grip upon our lives; a story sliding and stumbling over a range of meanings as wide as the mountains of Montana. Despised and feared as the darkest and most vile, adored and desired as the purest and most pristine, wilderness evokes love and hate. Ultimately, like any passion, it deals with the place of the other, and thus with the place of ourselves.

Initially wilderness was synonymous with threat, a hostility to be mastered. As a by-product of Western culture, especially agriculture, wilderness stood for that which was not-(yet)-cultivated, the dark and dangerous, the unlimited, lurking at the limits, always on the verge of overgrowing the fragile new structures of culture, be it crops or lawns, with their respective outcasts of weeds, witches, and (were)wolves: their gleaming green eyes staring holes into the walls of our cities and souls. But by the time European forests were clear-cut, the last frontier was won, the mysterious and tropical jungles were divided into profitable colonies, and the different outlaws and savages were assigned their proper places in mental institutions, hospitals, prisons, zoos, and anthropology books, the Big Bad Wolf reappeared, disguised as the desire of Little Red Riding Hood. Caged, trapped, burnt—devil set on fire—the wolf keeps coming back to us, because we keep coming back to the wolf, the hungry shadow of our imagination, smoldering in our soul.

With the woods and its denizens, such as wolves and bears, vanishing, the story of wilderness drifts into its next episode: fairy tales. The stories that spoke of a life *with* these others were on the verge of extinction, tales told by lisping old women living at invisible borders between villages and fields, with nobody listening to them anymore except some eager assistants of enterprising philologists such as the Brothers Grimm. Their stories turned into fairy tales, collected and preserved in the Big Book like dried flowers in the herbarium, pinned specimens of butterflies behind glass, neatly separated, ordered, and systematized: objects for study and research. Separated, they are silenced as living participants that carry their *own* determination in togetherness *with* others. In their second lives they do not themselves move anymore but are determined, either by categories of sciences such as linguistics, botany, and biology, or by the symbols of a diffuse but extensive realm of cultural imagination. While the hide of the last wolf fades away above the fireplace, father reads to the children before they go to sleep about the Wolf and the Seven Kids. They take the wolf with them in their dreams, the only place left in which to meet him. And thus in the middle of a cold Russian night the window of the bedroom of Freud's Wolf-man flies open and in the leafless tree, dark against the white snow, are seven white wolves staring at the little boy, their ears attentive, listening for something already there and about to come: screaming, the child wakes up, responding to the wolves' ears that already knew, already heard, his fear.[2]

The disappearance of European wilderness provided the young American nation with a new identity. Struggling with a cultural inferiority complex vis-à-vis Europe, the United States found a sense of self in celebrating its still pristine, "virginal" wilderness areas. Wild landscapes took the place of the missing national and cultural history. Congress officially acknowledged this in 1847 by acquiring Moran's majestic painting of the Grand Canyon for its Senate Lobby. The Hudson River School (1825-1875) incorporated it nationally and internationally into an American artistic identity by means of sublime imagery of the wild American landscape.

Ironically, the same period saw the most cruel impingement on American wilderness: between 1850 and 1880 more than 75 million buffalos were killed, mainly for their hides and tongues. Packs of wolves, following the American Fur Trading Company to scavenge the countless carcasses the carnage of the buffalo hunters left behind, formed an easy target for the same hunters. It was the livestock industry, however, that meant the end of the wolves of the plains. Cattlemen conquering the West replaced the indigenous grazers of the prairies with their own animals and hired commercial wolfers to take care of the wolves, who, losing their original prey, had turned more and more to the domestic stock. Strychnine-laced meat was spread all over the ground and the poisonous saliva of dying wolves, foxes, and coyotes dried in the grass and was stored for months—years—killing ponies, buffalos, and antelopes that fed on the prairie. If one includes the passenger pigeons that were used for target practice, it is conceivable that between 1850 and 1900 some 500 million animals were slaughtered on the plains. "Perhaps 1 million wolves; 2 million. The numbers no longer have meaning."[3]

Just as the cattle industry took care of the vast interior grasslands, the mining companies took care of the mountains and the timber industry took care of the trees. All of this left a legacy of exploitation and destruction on a scale whose effects are still not gauged fully, and probably never will be. The result, however, has come to be understood and legitimized as "economic necessity."

The only mystery, again, is the wolf. His eradication was so far beyond what was necessary, and the way in which it was done so rich in cruelty and perversion, that one can speak of a true holocaust in the sense of a totality of destruction—the Greek *holokauston*, "that which is completely burnt." Burnt sacrifice. By the end of the nineteenth century the wolf had virtually vanished from America. Poisoned, trapped, maimed, tortured, fed ground-up glass, set on fire.

The myth of the Wild West speaks with a forked tongue. The ideal of wilderness arose when we no longer knew how to live with the wild. From its beginning wilderness has been its own abstraction, sacrificing wildness on its altar.

II

The conceptualization of wilderness goes hand in hand with a reduced capacity for living *with* the wild, this otherness that is not controlled by human culture. The other becomes a symbol, no longer experienced as part of one's own life but externalized, categorized, stigmatized, confined to a determined area or to an abstract definition. There is no possible participation with wilderness as externality: either one conquers it, subsuming it under laws of civilization, or one occasionally goes out to it on vacation, as a retreat. Lévy-Bruhl contrasts this typically modern experience of compartmentalization and exclusion with the immersion intrinsic to primitive society:

> For the primitive mentality *to be is to participate.* It does not represent to itself things whose existence it conceives without bringing in elements other than the things themselves. They are what they are by virtue of participation.[4]

In his *Notebooks* Lévy-Bruhl makes clear that the qualification "primitive culture" does not imply an evolutionist hypothesis in which civilizations traverse certain stages such that our present phase is completely distinct from the "primitive" one. As he says, "This mentality called primitive . . . we constantly find around us, and even in us."[5] Even if the ways in which participation manifests itself differ considerably, "the root remains the same: the affective category of the supernatural."[6]

The language of Western metaphysics, however, is that of concepts instead of affects, images, or elements; it talks in terms of Being and Things instead of trees, water, the sky, the cold. The notion of wilderness has fallen prey to a metaphysics of presence and creation; rocks and winds have metaphysical power only insofar as they refer to the creator, who is modeled after "man, the maker," conceiving his world in clear and distinct entities or abstract definitions.

Whereas "wilderness" as a collective noun has been turned into a steady state with well-defined boundaries, "wildness" is "by definition . . . intractable to definition, is indefinite";[7] it signifies what is not determined and not easily grasped, much like mist lingering in the landscape. The wild is not confined and does not confine, just as the silent does not define and is not defined; they traverse the classical metaphysical distinctions between activity and passivity, subject and object, selfsame and otherness. The deer takes up the forest as much as the forest takes up the deer. These implicit relations resonate in Lévy-Bruhl's notion of participation and are ultimately constitutive for the possibility of individuality:

> If participation were not established, already real, the individuals would not exist. Thus the question is not: here are objects, individuals, how can they participate with each other? . . . but rather how some clearly defined individuals . . . disengage themselves from these participations.[8]

Lévy-Bruhl sees this question answered in modern society by the "substitution little by little of the affective with the logical abstraction."[9] This growing importance of a conceptual ordering finds its material expression in the rise of bureaucratic structures of an increasingly compartmentalized society.

III

Modernity is still characterized most forcefully by Marx and Engels:

> All fixed, fast-frozen relations, with their train of ancient and venerable prejudices and opinions, are swept away, all new-formed ones become antiquated before they can ossify. All that is solid melts into air.[10]

The very predicament of disintegration, tied to the ebbs and flows of the market economy, elicited a fierce control in the form of rigid bureaucratic systems on many different levels such as schools, hospitals, prisons, families, mental institutions. Ironically, this very effort to freeze fleeting social structures in a rigorous ordering of public and private life evinced more than ever a desire to escape. The modern era witnesses at the same time a firm consolidation of epistemological and ontological borders of what counts as "health," "sanity," "justice," and "knowledge," and a desperate search for ways out of these confinements.

The upsurge of the idea of wilderness and of the unconscious and the stress on sexuality should be seen in this light. The rise of the concept of wilderness coincides with the first serious thematization of the unconscious, the mental equivalent of wild land, and an obsession with sexuality, its bodily equivalent. Since modernity is fundamentally a patriarchal development, the main focus of the thematization of sexuality is on woman's body. Woman, like wilderness, figures as an abstract ideal who channels profoundly ambivalent emotions: as the mysterious other, she is uncontrollable but utterly desirable, and at the same time she is an other to be conquered and domesticated. Instead of a participant in life, she is an emblem of it; for Nietzsche, the prototypical Romantic philosopher, she is the abyss, the veiled one, the Truth. Breaking out of this abstracted paradigm means all too often becoming unintelligible and undesirable as a woman: many women writers or scientists have been unmarried, and there has been a high rate of hysteria or suicide among the gifted sisters of famous men. Similarly, when nature breaks out of the imagery of wilderness, it loses its imputed magic and is reduced to a mere resource, rather than being a regenerative source. Woman and the wild, thus hypostasized in static concepts, reduce reason's fear of disappearing in its own longing for its opposite, but leave the desire unfulfilled; an unfulfillment that makes the force of the other—the mystery of animality, wildness, and bodily desire—only stronger. Hence consciousness finds itself in a double bind, captive of its own mind, uncomfortable in its own flesh.

This is why Dostoyevsky's underground man laments his "unfortunate nineteenth century":

> I would now like to tell you, gentlemen, whether you do or do not wish to hear it, why I never managed to become even an insect. . . . I was not deemed worthy even of that. I swear to you, gentlemen, that to be conscious is a sickness, a real thorough sickness.[11]

Caught in clear-cut conceptualizations, "safely" locked behind the doors of private homes, the gates of parks, the boundaries of reason, woman, wilderness, and the unconscious are doomed to function as *others* that do not participate in life but are lived out in the moods and yearnings of pale and distraught gentlemen.

IV

It is time for breaking (with) the concepts of wilderness, woman, sexuality, and the unconscious, granting (as a paradoxical but strategic interlude) their material effectiveness. Initially we might even fight for the expansion of the odd areas designated as "wilderness" and for the acknowledgment of women, savor other modes of consciousness, or liberate sexuality. But at the same time we should not forget that what is ultimately at stake in this momentary but perhaps necessary essentialism is the place of otherness.

Rethinking otherness requires understanding the role the other occupies. The very externalization of the other subtends the modern notion of subjectivity, grounded as it is in a sense of lack. The modern subject is a subject of desire, the desire to become subject vis-à-vis the postulation of an object. A movement in which both are frozen into static identities, "fixed," as Irigaray says, "not 'free as the wind,'" because this subject "already knows its object and controls its relations with the world and with others" instead of living with them.[12] However,

> any *finalization* of identities is anathema to otherness. . . . This is why wildness, which contradicts any finalization in identification, is at the heart of otherness, as well of course at the heart of any *living* self or society.[13]

Wildness is as vital for the self as for the other. It prevents any closure of communities with its imminent danger of stigmatizing the other into Jew, woman, wolf, or woods. Wildness guarantees the openness of any being in common, the sharing of commonality by different entities in a common place or project. As soon as this commonality closes its borders, that is, eradicates its wildness, it is reified into a "common Being" of a more or less totalitarian cast.[14] Wildness, so easily stigmatized itself as "something out there in nature," may be crucial in a rethinking of contemporary forms of sociality. Preserving wildness ultimately implies a political and ethical commitment to resisting any encapsulation or segregation of otherness.

The question is, how do we preserve this kind of wildness, or, for that matter, retrieve otherness from its isolation within a merely symbolic status?

An acute sensitivity to sense-experience with its unremitting attention to the specificities of other bodies, be they human, animal, spatial, or elemental, is one mode of opening up the here and now of everyday life into more unpredictable, wilder, layers of experience. A delight in a heightened sensitivity of the senses, which we might see as a mode of sensual pleasure, can reopen and reverse this [modern] conception and construction of the world. It can return to the evanescence of subject and object. To the lifting of all schemas by which the other is defined.[15]

Such a receptivity for sense-experience creates a familiarity with the intricacies of the other or the particularities of the places we inhabit—in the face of which any static conceptualization would be undermined—so as to make room for an awareness of a never-conclusive set of performative possibilities of the other, a richness of performative behavior that manifests itself implicitly through complex patterns of adaptive and generative activities such as we see in wolves roaming *through* the land, *through* the seasons. Rick Bass invokes this performativity when he points out that "the thing that defines a wolf more than anything—better than DNA, better than fur, teeth, green eyes, better than even the low mournful howl—is the way it *travels*."[16]

Lingering in the journey implies porous conceptual borders, which dissolve in the complexity of different modes of participation in the landscape. Only when there is an openness to the experience, that is, a participation in specific activities and places, can we keep the wildness of the other even in wilderness areas. Wilderness, then, "is not only a place you go. Wilderness is what happens to you. *Shivered, sweated . . .*"[17]

In participation things happen to you. Walking the strenuous gradient of a steep slope, the rhythm of the mountain pulses through your blood, and your body sweats the heat of the day until night comes in and cold covers the rocks, the trees, your bones. Wildness pervades us if we are open to it and participate in it. It is implicit in us and we in it.

The unnamed and untamed slip between borders into the "darkest woods, the thickest and most interminable swamp," or "impermeable, unfathomable bog," escaping human measurements of words, definitions, numbers, or buildings and roads.[18] The wild is "refreshing" or re-creating precisely because it is never fixed but always transcends the frame of our thoughts. The unframed is the place where our thoughts and our desires can roam freely, joining that which is already unfathomable: nobody knows where a mountain begins, or where the sky ends. It is a tacit knowing that is always to be determined along the way, never caught in stable definitions but always moving. Something implicit.

Neither defining nor confining, the silent and wild leave open a space for wandering and wondering; it is where our clear and distinct grasp transcends itself in impermeability, where our knowledge and being appear in the mode of dis-appearing. As soon as one stakes out the wild, it is gone. As long as one can be a presence that leads to its own dis-appearance, one stays wild. Disappearing not in the sense of simply vanishing from existence, but as a beginning *not-to-appear,* a not standing out anymore.

Rain waiting in the trees, branches dark, wet, and soaked through with water, clear crystals hanging down, carrying the whole world shining and upside down . . . all gone when you grasp it, leaving the branches dark and alone with their shapes. We have forgotten how to vanish; too much of what our words and hands grasp stays present, sticks to our future, covers our past. We stand out but have lost our capacity to disappear. And therefore so much around us and ultimately so much of ourselves has in fact disappeared, or has been drained, clear cut, paved over, eradicated, eroded.

Forgetting to disappear means not knowing how to stand out. Not knowing how to die means not knowing how to live. Not knowing how to be silent is not being able to listen or speak. Standing out, living, listening, speaking is a question of respecting the porosity of limits. In sheer standing out, identity fixates itself as well as "its" otherness. Standing out in disappearance, however, is making room for the other to appear: the foreground acknowledges the background by permanently shifting ground.

V

Wilderness is not silence. Silence does not exist. In complete isolation John Cage hears his blood flowing and his nerve cells firing.[19] Silence is not wilderness. It is time to leave wilderness behind us. It has had its function. Now is the time for the specific, the here and the now. Wild is the scent of your sweat, of salt, of sea. The mountain without roads. Wild is: No Roads.

There are different modes of change. Things, including living things, have duration, they exist over time, they follow the heat and cold of the seasons, becoming warmer and colder themselves. Not so the cold itself; becoming warmer, it will disappear in the warmth and is as such more versatile, more fluid, than any of the things through which it manifests itself. Only the snow man is true to the cold: along with the vanishing winter he loses his substantive presence in an indeterminate pool of water saturating the thawing earth. To him, in his quality of nonadherence, Wallace Stevens dedicates the poem "The Snow Man." For him, who does not speak, who does not name, but listens, Stevens let the winter wind blow with its sound of a bare place. The snow man is no-thing and hence beholds nothing. Silently he stands and listens until he hears the spring coming with soft winds, and green grass shines through his disappearance.[20]

Elusive presences. Brown velvet wings, a butterfly lands on my knee, explores with its tongue my skin. Instinctively I whisper "Hey," as if welcoming a friend. Disturbed, it flaps away and lands quietly on a flower. I feel inadequate: I do not know how to greet a butterfly in the spring.

A warm summer evening. I spread my sleeping bag out on a soft patch of moss and sit down on an old tree stump enjoying the quiet evening. Suddenly leaves rustle and a big raccoon wobbles out of the bushes. Agile with its little hands, it inspects my camping spot and pulls the pad from underneath my sleeping bag. I grin. The raccoon looks up; I try to make it feel comfortable again with some soft sounds that usually calm my cat and dog. In a flash the animal disappears in the woods, leaving me alone with my lack of subtlety, my inadequacy in knowing how to speak to a raccoon.

Only when I am silent do those animals take me up in their presence; only when I stand out *and* disappear at the same moment, can they be with me. Only when I am just there, without being present too much, do they come back.

As with seasons, the being of the wild is change; it is a fluttering companion located between waking and dreaming; a colorful connection between knowing and not knowing; never caught in rigidity, it always moves, comes and goes.

VI

Silence. The very absence of content, or for that matter, of *telos*, opens up the possibility for other things to appear, to show themselves. Silence opens

> the doors of the music to the sounds that happen to be in the environment. This openness exists in the fields of modern sculpture and architecture. The glass houses of Mies van der Rohe reflect their environment, presenting to the eye the images of clouds, trees, or grass, according to the situation.[21]

This is an openness that phenomenology tries to retrieve by stepping back to show what is already there. Stepping back in order to let the here enter in its fullness. Without prejudices, without expectations.

> We must consider speech before it is spoken, the background of silence which does not cease to surround it and without which it would say nothing. . . . We must uncover the threads of silence that speech is mixed together with.[22]

We see here a structural equivalence between silence and Lévy-Bruhl's idea of participation; both function as conditions of possibilities instead of as determined entities. The voice of silence opens to different voices, its pace to different rhythms; its character or definition is formed by the possibility of receiving different characterizations.

Silent is Santayana's landscape. As one of the few philosophers in the previous century who thought landscape worthy of philosophical reflection, he considered its primary trait to be *indeterminacy:* "A landscape to be seen has to be composed." The ability to appreciate wilderness landscapes he called the "mastery of the formless."[23]

Silent is the stream's roaring path through the forest, not because there is no sound but because the water's speaking is immediate, unmediated, without representation. The river does not name. Silence is not-naming, it is letting things appear without interpreting, translating, or casting in static forms; silence affords a place for many sounds. Indiscriminately the river takes up what comes along and has its say by washing away.

Silence is the "resting place," where the attention given to words, an attention that has stolen the secret of reminiscences, can evaporate: "This secret is only the inner presence, silence, unfathomable and naked."[24] Silence slips between representing something and being the something that is represented.

"The wild does not have words. . . . Its unwritten pages spread themselves out in all directions!"[25] In February winter was settled on the Island. The salty Sound was frozen, its waves still in silent ice, firmly held by the cold. Just by the passing of time, of season, a strip of wilderness has emerged all along the northern coast of Long Island. Water has turned solid—an elemental change—and we can walk over it. Jacob, the dog, immediately rises to the occasion: agile as a polar hound he inspects the newly formed arctic tundra, chasing screaming seagulls over white slates of snowy ice. They leave him alone with their loud laughter, resonating sarcastically in the thin air, while their white bodies circle safely away in the blue sky . . . only to return with the geese who carry warm winds under their wings, winds that blow the ice into floes and disperse the Nordic wild over the sea to be taken up by the water again.

Wild is what comes and goes; a flock of geese, Heraclitean flux. Fog in the early morning, creating a thin slice of wilderness, land of mist, of mystery, of layers; the intimacy of a quiet time before the rush-hour traffic hits the road.

The silent and wild appear *and* disappear, ever evading closed definitions. As soon as they are defined into a special name, you have to be on your guard, because wild animals just appear, like the butterfly, the raccoon; or we follow their disappearing tracks, through snow, bushes, and deserts, over the mountains. They don't come when we call them as does Jacob, or Monday the cat. When we call wild animals by a name they disappear. They vanish in the bushes, are taken away by the wind, sometimes forever.

Nondomesticated animals such as birds are just around, a general presence, a sound in the air. When we see them regularly around the house we begin to recognize particular ones, like the reckless Carolina wren who eats Monday's cat food from her bowl while she is snoozing next to it, or the blue jay couple that nested in our garage. Being in our life, they are in our language: "Did you see that the blue jays' eggs are hatched?" We talk about them, but rarely give them a personal name. They are birds. They are around: in the trees, in the air.

Yet by the time they are extinct, they have become distinct, they have a name, they have become true individuals, in-divisible. No further division is possible anymore, they are single specimens. They don't fly around anymore, they don't sing anymore, they are singled out. Caged and named, they stand before us, and, after attempting special breeding programs, we all too often have to witness their death.

Martha was the last passenger pigeon; she died in 1913 in the Cincinnati Zoo. Lady Jane and Incus were the last two Carolina parakeets. This native parrot of the United States (the only one) was a "pest" for agriculture and thus was doomed to be wiped out; the last two died in a zoo in 1918. Orange, the last dusky seaside sparrow, died old and infertile in the eternal youth of Disney World; he was driven from the air to make room for Cape Canaveral space explorations.

Named, they lost their time, their place: their last days were spent out of place, in zoos and amusement parks. Martha, Incus and Lady Jane, Orange.

VII

To be wild is to stand out *and* to disappear. But we are far from that capacity to disappear, to be silent. It is hard for us to pause. Therefore, we need to limit ourselves and subject ourselves to laws, to a wilderness act that states explicitly:

> A wilderness, in contrast with those areas where man and his own works dominate the landscape, is hereby recognized as an area where the earth and its community of life are untrammeled by man, where man himself is a visitor who does not remain.[26]

So we can give ourselves time to learn again to stand out and to disappear in the intimacy and participation with the community of the land.

> Intimate land:
> > The canyon
> is a silence
> in the center of a Self
> weightless as the silence of a rock,
> a silence that is felt
> > pretending you're not there,
> > sitting still with stones
> > to be weathered by the day,
> > > forgetting
> who you are[27]

Intimacy is experiencing something together, a *shared* grammar, forgetting who you are. Intimacy is not a *declaration* of love that is spoken by me to you; it is not just inside me, nor inside you, but in and between us, most of the time not even in words but enacted, the way animals love, or the land takes us in. Monday the cat, arching her body to find the hollow of my hand, caresses in one and the same move me and herself. The caress creates an "us," a moment in which we belong together, are part of each other, participate in each other. Me, you, the cat, the canyon. Not a participation in the sense of relation between two separated entities but an implicit connection in which individuality at once arises and dissolves, as described by Lévy-Bruhl.

Wild is what travels through our skin, through our borders. Clouds roaming the sky, worms slowly inching through heavy earth, way under any property lines. Entering our borders, however, means all too often the end of the wild. Road kills take care of most of America's wildlife. But borders go far beyond interstates, property fences, iron bars of the cages in the zoo. The most dangerous and insidious ones are the implicit ones, the vast invisible borders we trace before we know: the cyanide we use to separate gold from the mountain slowly seeps into the groundwater, evaporates, and condenses into acid rain, pouring down hundreds of miles away for years to come. A sweeping movement of unintended borders, spreading wildly and out of control, through time and through places. The clear-cut mountain not only lost its trees but everything that participated with the trees, from spotted owls flying high above, to fungi and worms deeply buried beneath the roots. All gone. Forever. No soil left to leave one's tracks in. All disappeared, including the regenerative capacity of appearing again.

For a while, maybe, we did not know how our borders go underground or dig through the sky only to show up unexpectedly but disastrously. Now we do. Now we know that they don't stop at the surface of a particular enterprise, but that they have their own life, eating their ways through mountains and oceans, leaving behind trails devoid of life.

VIII

Again and again the same story. As soon as one of the few remaining, shyly returning wolves eats something that "belongs" to us like a pet or some cattle, the outrage is complete and the witch hunt starts again. Rick Bass follows this flame of irrational hate in his gripping account of the Ninemile Wolves in Montana:

> What was lost, in this whole story—two steers, and two lambs? May we all never be judged by anything so harshly or held to as strict a life or unremitting of borders as the ones we try to place on and around wolves.[28]

What is judged here is the wild, the other, culminating in this all too charismatic symbol, the wolf, whose green eyes lock us into superstition. Wolves do not need us. The wild does not need us. It does need us to be capable of disappearing. We need the wild, our body running, loving, the storm at sea, the geese screaming and flying on their migrating path, the gentle sound of spring rain on new leaves. We need the wild, the other, to enhance and intensify our everyday life, to allow us to experience that there is always more to the here, and that this more is always already *here*. Here in our body, in our thoughts, in our lives, in the geese, in the ground under our feet, in the boulders lying at the beach since glacial times. In the howl of the wolf heard by the Montana wind. The Rocky Mountain gray wolf, *Canis lupus irremotus*, "The Wolf Who Is Always Showing Up,"[29] appears to disappear, only to appear again.

Notes

For his many helpful suggestions I wish to thank Ed Casey.

1. Edmond Jabès, *The Book of Questions, Yaël, Elya, Aely*, trans. Rosemary Waldrop (Middletown: Wesleyan University Press, 1983), 26.

2. Sigmund Freud, "The Occurrence in Dreams of Material from Fairy-Tales" [1913], in *Collected Papers*, vol. 4, trans. Joan Riviere (New York: Basic Books, 1959), 236-43.

3. See Part Three, "The Beast of Waste and Desolation," especially Chapter Nine, "An American Pogrom," in Barry Lopez, *Of Wolves and Men* (New York: Charles Scribner's Sons, 1978), 167-99.

4. Lucien Lévy-Bruhl, *The Notebooks on Primitive Mentality* [1938-39], trans. Peter Rivière (Oxford: Basil Blackwell, 1975), 18. Emphasis mine.

5. Ibid., 142.

6. Ibid., 143.

7. Thomas Birch, "The Incarceration of Wildness: Wilderness Areas as Prisons," in *Environmental Ethics* 12 (1990): 3-27, quotation on 8.

8. Lévy-Bruhl, *Notebooks*, 18.

9. Ibid.

10. Karl Marx and Friedrich Engels, "Manifesto of the Communist Party [1848]," in: *The Marx-Engels Reader*, ed. Robert C. Tucker (New York: Norton, 1978), 476.

11. Fyodor Dostoyevsky, *Notes from Underground* [1864], trans. Richard Pevear and Larissa Volokhonsky (New York: Knopf, 1993), 6.

12. Luce Irigaray, *An Ethics of Sexual Difference*, trans. Carolyn Burke and Gillian C. Gill (Ithaca, N.Y.: Cornell University Press, 1993), 185.

13. Birch, "Incarceration of Wildness," 11. Emphasis in the original.

14. See Jean-Luc Nancy, *The Inoperative Community*, ed. Peter Connor (Minneapolis: University of Minnesota Press, 1991), especially 1-43. See also Giorgio Agamben, *The Coming Community*, trans. Michael Hardt (Minneapolis: University of Minnesota Press, 1993).

15. Irigaray, *Ethics of Sexual Difference*, 185.

16. Rick Bass, *The Ninemile Wolves* (New York: Ballantine Books, 1992), 6. Emphasis added.

17. David Strong, *Crazy Mountains: Leading from Wilderness to Turn Technology* (Albany: State University of New York Press, forthcoming, 1995). Emphasis in the original.

18. Thoreau, "Walking," in *Walden and Other Writings of Henry David Thoreau*, ed. Brooks Atkinson (New York: Modern Library, 1950), 615-17.

19. John Cage, "Experimental Music" [1957], in *Silence* (Middletown, Conn.: Wesleyan University Press, 1973), 8. Cage adds: "Until I die there will be sounds. And they will continue following my death. One need not fear about the future of music."

20. Wallace Stevens, "The Snow Man," in *The Palm at the End of the Mind*, ed. Holly Stevens (New York: Vintage Books, 1972), 54.

21. Cage, *Silence*, 7-8.

22. Maurice Merleau-Ponty, *Signs*, trans. Richard C. McCleary (Chicago: Northwestern University Press, 1964), 46-47.

23. George Santayana, *The Sense of Beauty: Being the Outlines of Aesthetic Theory* [1896], ed. W. Holzberger and H. Saatkamp, Jr. (Cambridge: MIT Press, 1988), 85.

24. Georges Bataille, *Inner Experience*, trans. Leslie Anne Boldt (Albany: State University of New York Press, 1988), 16.

25. Tomas Tranströmer, *Selected Poems 1954-1986*, ed. Robert Hass (New York: Ecco Press, 1987), 159.

26. Part of Section 2 of the Wilderness Act of 1964.

27. Kirk Gittings and V. B. Price, *Chaco Body* (Albuquerque: Artspace Press, 1991).

28. Bass, *The Ninemile Wolves*, ix.

29. Ibid., 14.

9 . The Idea of the North: An Iceberg History

David Rothenberg

The idea of the North marks a place for the idea of the wild. The idea of the wild is at first something to fear, a dark trait inside us that fights back before civilization emerges. It is what propels the way of life thought to be nasty, brutish, and short, a nature we are supposed to be happy to have escaped from. We remember it still as a latent possibility, lurking inside us, remaining untamed.

Feared for years, centuries, epochs, this wilderness at the edge of all maps, the country of the unknown. Cultures grow, expand, and conquer, based on the need to fill up these white spaces. As we succeed, the tangibility of the emptiness recedes. It is farther and farther away. East, and West, unexpected oceans, unprecedented continents. Magellan's ship made it around the world; suddenly there is less mystery.

But in the North, it is still cold. Few consent to live and inhabit there. It used to be warmer, back then in the days of the Viking pillagers and adventurers, with their *grønland* and *vinland* named to entice others out from the comfort of their homes. They risked death and worse, loneliness, to yearn for the unknown. Now it has frozen down, and just the intrepid care to remain in these places—those who need the crisp silence to think and to exist.

It is only once our now-global culture senses the glimmering of limitation that the wild is craved as solution or antidote. Out of Romanticism comes respect for the wild, in opposition to the machinations of an increasingly planned and ordered system of labor, punishment, and reward. The wild turns this off for an instant, returning us to instinct. And even now the North retains an iota of this inscrutable wildness.

And no one will tell us when we have reached there, as it still is an idea before it is fact. I sketched some of these ideas with words when I once visited and imagined how far I had to go. Their rhythm sings from this other wild place, and they appear in italics throughout the text that follows:

you won't know when the North is reached
because wherever one alights
you will meet people coming
from a point just farther north,

and they will smugly survey the surroundings, telling
of the beauty back up in their NORTH
a place beyond even their home
in the cold windarkness and stone,
by edges, no, in water and yes, ice,
that remarkable and frozen land . . .

and they may not savor what appears around them.

"No," they whisper, "this here is not the North."
"You must come north with us."
Yet as you continue traveling up
toward circles of tundra and rime
satisfaction will never come.

because there will always be someone
to tell you that you have so much

further to go before you reach the North.[1]

And this is only the beginning: We still want it, we still follow the compass and travel some more. Icebergs floating, icebergs of thought. Stories recalled, a catalog of northern whiteness, carried by the sea, scraping ships, breaking up against one another. This is the way the ideas of North emerge: Across the gray-blue horizon, nothing is moving. Then a piece of white peeks out of the waves. It is nothing, it's harmless, just a bright glimpse. But beneath the dark surface is a mass of sharp danger. Still, we remember first the light, and the fullness of sky.

This northerly want, a satisfaction within open spaces, an exact science left out so we can fill it with our purest ideas. Few can be happy there, covering all extremity, where little remains to challenge the most human of visions. There may not be much pungence, or thick jungle air, or even the bustle of human community. But there is the space for ideas, the need to amuse, the tabula rasa of always frozen solid ground. It is the realm of ideas guarded or surrounded by space, and home to a special aesthetic that has found at least one way to solve the problem of the wild.

What is the problem of the wild? Not the nostalgic yearning for a better past. Not at all. It's knowing that an affront to culture and order lurks always inside us, resisting all plans. The wild wants to break up our marriages, tear down our buildings, dissolve our corporations. It says no to our control. Yet without it we die. It keeps us wanting, forcing the obscurity of the path we must take.

The North answers this with clarity and with space. To civilization, it reflects a cool indifference. Our skies are still greater than you. Our nights are alive with sparks and a glow that you will not equal from below. There is, you see, no reason for it to happen.

You hear the word North,
you then think of the light.
It's the light they've told you about,
the special light,
the glowing yet soft light,
the crisp yet dim burn
of everevenings.

That's what they tell you, the light:
so marvelous, such difference.

But it's the darkness they mean.
this is what they want to
tell you about
but they cannot. It is forbidden

It does not open and enable the heart
as talks of light and golden shadows do.
Yes you know it
can be always light for a time, true,
those months are like years
of wonder, of freedom.
In those times nothing is hidden
one must always see all things at all times,
for they will not be shut out by the sun.

And the times, and time,
that is something in its constancy
that is with you always then—

but the darkness, well, with darkness
one has no need to think about time.
Time is overcome, it is gone,
one is there, waiting, for nothing more,
one knows that the sun will never come up.

So with that thought there is nothing there
but waiting, an ephemeral moment

                                                    of waiting.
. . . and the biting cold . . .

(the precocious dawn, always too soon!)

that with the darkness has something to do

So we wish overtly for those endless summer evenings, knowing they will never conclude with the seclusion of night. But not so secretly we all need the night, and cannot even dream of what we would do with a darkness that lasts uninterrupted for months. That is the simplification of winter up here, of course. It is more like a continuous soft twilight, punctuated by silent aurora dances in the hollow blue cloudless skies. It may be cold, but it is a crystalline chill, just enough light so you can see through it to forever.

The idea of North is with us from any latitude, close by any season and surrounding heat. We need it to imagine more lucidity than the boom of usual confusion admits. It's wilder up there, fewer people, more lynx tracks, caribou, traces of bear. On his deathbed Thoreau murmured "moose . . . Indians" as final words in this life; he was thinking not of his trip to Cape Cod, but of those long weeks in the interior of Maine, what he knew as the North, the immediate memory of wilderness always close to his heart, even at his last moments on earth. Perhaps afterward, right away or after a time, a return, back to the North, in an uncrowded place at last.

This aspect of the North is a location for that need for the wild, a free spirit always latent inside us but demanding of a geography to bring it to life. In the summer from the city we look to the North where it is cooler; in the winter it is the antidote to lukewarm winters, the last remaining place where it really does snow. Maybe it is just regional; you all may look to the West or the mountains or somewhere in any direction for the fulfillment of outer wildness, but of all those places I have found myself the North has promised fewer people, more solid nature. Or more horizon to imagine something ineffable we want to call nature, wishing it were a tangible, huge, and surrounding thing.

Mark Helprin's wonderful novel *Winter's Tale* takes place in two times, one hundred years ago and today. In the old days everything was, as we often dream it, larger than life, with the winters colder than cold, snow billowing through the streets piling in drifts above the doors and windows above our heads. The characters travel from New York City north up the Hudson River on swift horse-drawn sleds over the ice, far into the interior northern regions of the state, to the mythical Lake of the Coheeries, where the tale of winter never ends, where the snow and ice are permanent but not dark—instead, joyous and exhilarating. Culture is not lacking, but is self-contained. The relatives living up there are brilliant at the kind of knowledge that can be sequestered and learned in isolation. There is one, I remember, who is a master of words, a linguist beyond compare, who has memorized dictionaries in those long winter nights and who finds no language puzzle to be in the least bit difficult. Sheltered, holed up for the winter, needing to speak and to read. Much to learn, the construction of vast and private worlds. There is time to think in the North, and the sparseness can be filled as easily with density as with singular truths and sharp insights.

I have lived in northern countries, studied their languages, noticed how few words can make do for many. Meanings overlap, and strange parallels chart the culture whose words stand for more than one thing so that only context can resolve the confusion:

> *the word for crowd is the same as distress*
> *the word for reason is the same as abyss*
> *the word for teach is the same as to learn*
> *marriage = poison, goddess = haze*
> *similarities only outsiders notice*
> *that's how we shape characters,*
> *and discover what belongs with who*

Knut Hamsun knew this sparseness of northern language well. In his *Pan*, Lieutenant Glahn glimpses women real and imaginary, appearing before him at his isolated outposts. Communication between him and them is always oblique. The message never gets through. In the end he is left alone between the sparseness of words. Travelers, I guess, find what they want up there. There is the choice of the individual involved alone, the encyclopedic, or the sparse way of talking that catches the contours of the land. Anyone who's been up there has heard similar spaces between words:

> *"Do you think I'm cold?" she said.*
> *"No more than the sky," he said.*
>
> *We sat looking directly at each other.*
>
> *behind were opposing windows*
>   *"it is so gray," said he,*
>   *looming behind the glass a dark hue—*
>   *pallid, terse*
>
> *pulling her closer*
>   *"what do you mean?" she said.*
>   *"look outside."*
>   *"no . . . you look outside."*
>   *"what— . . . Oh!"*
>
> *he turned around and saw her view, a molten orange,*
> *an alpenglow, volcanos erupting violet over castles,*
> *swirling penumbral fires, flare cast off as magnifications . . .*

"Oh, that's not my view at all"

the same world?—to be satisfied, never turn.

"Talk more . . . say something to me . . . whisper; I can't sleep."
"I thought I talked too much."
"No . . . not now."

In these places you hope to be with someone who silently smiles a knowledge that you both know; discovery of a secret North, unknown, unvisited places where the people who live there simply must stay. They have their own inner culture, a habitation of space. Words mean order; they must be held back. They have powers, and have been around a long time.

> The condition of life in the Far North still approaches the experience of the hunter-gatherer world, the kind of world that was not just the cradle but the young adulthood of humanity. The North still has a wild community, in most of its numbers, intact. There is a relatively small group of hardy individuals who live as hunters and foragers and who have learned to move with the mindful intensity that is basic to elder human experience. It is not the "frontier" but the last of the Pleistocene in all its glory of salmon, bear, caribou, deer, ducks and geese, whales and walruses, and moose. It will not, of course, last much longer.[2]

> What has become of our adventures—the ones that led us over icy passes, across great dunes, or often simply down highways? . . . We were not content simply to go adventuring. We also made our adventures public, in reports and accounts, profusely illustrated and secretly fostering an illusion: that even the most remote and distant spot is as accessible as an amusement park, a twinkling Coney Island; that the world has grown smaller with the rapid development of our modes of transportation. . . . But that is a mistake! Distances remain as immense as before. The line of flight is only a line, not a road. Pedestrians and hunters we remain.[3]

Do we go to find this remaining community, or to make the distant and frightening spaciousness seem close at hand and comprehensible? Every time we hear a story of anywhere so far away, we choose one of these attitudes toward it. It's either a report of a way of life apart from us, or the recounting of an adventure. If the account is truly adventurous, it may draw us further to the possibility of a community in otherness, a way of life that embraces space without trying too hard to fill it. What follows are some of these stories.

Knud Rasmussen, the Danish explorer, alights on the coast of Thule, western Greenland, to investigate the northern country. The explorer is supposed to conquer, to give names to what outsiders do not know. But Rasmussen is different from most. He wants to hear what the people have to say, to ask for their own visions of their own history. He looks for anyone who might remember the stories of a lucid past, when the impossible may once have happened.

"Me? I'm just an ordinary old woman," says Nalungiaq, from under a thick fur coat. There is a photograph of her staring sternly from beneath a sealskin hood, gazing directly at the horizon, pondering the visitor's request: "I haven't got much of anything to tell you." She hesitates a moment and then stares far across the snowscape and into the wind. "Well, I could tell you what life was like when the world began."

"Yes," says Knud, weathered but patient after years of searching for stories just like this one. "What was it like?" Her answer hints at how vast the power of human language might be, if this language might speak with the world, rather than define it away into silence. It might be the best story he ever heard. Here are his remembered verses from the North:

> In the very earliest time,
> when both people and animals lived on earth,
> a person could become an animal, if she wanted,
> or an animal could be a human being.
> There was no difference:
> All spoke the same language.

And the dogs and wolves stop howling to one another for the moment as the words begin; hearing the meaning of the hush, they look up to Nalungiaq as she speaks. People and animals living together on earth. That time could be now. And what is it that we are meant to share? Why do we care? Realizing our humanity seems to deny something animal that always walks with us. We watch them all moving shadowy around us. We imagine they can speak to each other, leaving us out of the talk. At least they seem to sense their own space in the world, knowing territory and purpose from birth all the way to death. We point to them running from our recognition. They are, understandably, worried about us and what we might do to their world, immediately or far away at the limits. We've lost our place because we are poised to wonder why. Yet once we knew one another's ways, and lived together in the identical community.

Touched by the power of sound; people, earth, what lies in between. The visceral appeal need not have a reason underneath it. The flow of sounds has its own order, easy to love without knowing how it all works. This is one kind of magic; astonishment at happenings in the world without wanting to explain. (The fire dies down, the song continues, and the words repeated again and again lose their senses and seep into sound:)

> That was the time when words were like magic:
> A single word, spoken by chance,
> might have strange consequences.

Signifying syllables become the human mark, and here is a concise reminder of the greatest power language has: we say things, they come to pass. Perhaps Nalungiaq means words are magic when they start events in motion without knowing the outcome. And that's the way it is, now or then, before or after, past to future. There is no control, with messages sent one way, received another, cries turning into calls into melodies into changes in the land and time, no action without reaction, no plan without a void unchanged. No name reduces the unknown, only shoving it off to the side instead.

To make words like magic again, we forget what they mean. Say any one of them over and over again so that it is just a sound thrumming inside you, drilling its emptiness into your soul. Sentences as sounds precede their sense as purpose. It seems to be only the human beings who name things. The people. The place. The not-human. The voice. The change. Accident or design? The world is still the same and not the same. It is language that wants it to be one or the other. The name must hold. And the thing is gone as soon as we have identified it. Left with only the name, we hold empty words. These incantations and syllables swirl around the elusive subjects of reality. If only words could be more than words! Trap any one of them in a corridor of parallel mirrors, and the single name immediately echoes to infinity. Inside the scale of that sound is a region of magic, where what is awesome pushes through the confines of reason.

The ancient sounds of the animals were more, says Nalungiaq. They contained within themselves the unquestionability of the demands they announced. "Taoq! Taoq! Taoq!" says the fox, "Dark! Dark! Dark!," so he can hunt for his prey without discovery. "Light! Light! Light!" sings the rabbit, "Uvdloq! Uvdloq! Uvdloq!," so he can bolt away without being captured. Between them they change things. There is day and there is night. Up North one lasts but a few months and the other swallows up the rest of the year. It was a good fight. Or a better conversation. But if such phrases have power, you had best watch what you say. One single word was once enough to make a difference:

It would suddenly come alive
and what people wanted to happen
could happen—
all you had to do was say it.

So that's the secret to the power of language. Call those few words to life, and they blur the sense between the human and the animal, crossing the boundaries of what each species alone might understand. Just speak, and what you want may come true. That makes the time that is the first time real.

Wishful thinking? Sure, words put together can make things happen. That is the essence of their magic. Stories can be dreamt up, and then they may be built into the truth. But to be instantly and suddenly powerful enough to reinvent the world! Here's how it happens: somebody gives some part of nature a new name, and then everyone else begins to see that thing where no thing was before. The name lets us hold experience in our minds and memories. We can remember it and share in the vision.

Nobody could explain it:
That's the way it was.

A sad longing for a time we never did understand. Naming takes us far from the animal powers even as it brings us close. That's the way it still is. What is most important cannot yet be explained.

Rasmussen takes his time searching the coasts of Thule for the most magic of words. He scours the coast, sailing in and out of every bay. This is his fifth time on these nominally Danish but very foreign shores. It is 1925. He has long since ceased imagining any practical passage to ease navigation through the Arctic. By this time he has seen a tragedy rise at night with the northern lights: these are people with ancient stories that are fading. The value of the ebbing words will be worth more to outsiders than the practical resources of the land itself. The people he finds implore him to listen closely to the words, and to repeat them over and over again. These turn out to be the chains of syllables whose meaning no one can quite remember, with rumblings reaching out to the animal and spirit realms. There is no successful incantation out of context, so the repetition of phrases out of their context serves no one, neither the speakers who conjured the world once out of sound, nor the would-be explainers of the present day. If this breath from the Arctic can be felt down here in the warm, humid sun, it is because the icebound and wide-horizoned sky is still within reach. They are not so different from us then. So what if it turns out the people of the North have no more words for snow than we do! Each kind of white crystal that annually blankets their world means more to them with expectation than the freak blizzard that is strong enough to close the city down in a muffling, modern silence. To speak the world so that you connect to that world requires a very special kind of life.

See the circle traced by a blade of grass blown by the breeze at the beach, reiterating a narrow arc in the sand. That too, is part of the language. Receding water with the tide, ripples on earth as well as in ocean, no ridge exactly like any other. And on snowfields in the high mountains, melting each year unevenly but in a way familiar, pockets and peaks called suncups, looking like the ruffled pattern on tidal sands, only in snow. An earth language, which you could feel in both places, running your hand across the wet surfaces. From speech to the silence of seasons moving over time. Magic words? We can lie quiet with them as the shapes of the earth talk back with shadows.

Every adventure becomes another book. Every book, I hope, begets another adventure.

> If you choose, you could skip the pages into some Arctic lake one by one, and watch them . . . sink and lie shimmering among the greenish rocks; and the water rippled over them in the wind, as if trying to turn them, but they would never turn or be together again.—All books are like this; they stand shoulder to shoulder in the library stacks; perhaps they are "popular" at first, perhaps not, but eventually they stand anonymous, unread, forgotten; and that is how it should be, for that is how it is with lives.[4]

In 1993 a very strange mystery novel was translated from Danish into English, *Smilla's Sense of Snow* by Peter Høeg. The unlikely heroine is a half-Inuit glaciologist named Smilla Qaavigaaq Jasperson, a Nalungiaq of the modern world. A young boy falls to his death from the roof of her apartment building; from her years in the Arctic she senses footprints on the rooftop snow that indicate foul play. She decides to investigate the mystery; it becomes complex, unsettling, detailed yet somehow empty, involving her physician father, the government's systematic oppression of natives, international smuggling, and this independent woman's search for acceptable love. We do not learn enough of supporting characters; their lives and deeds make little sense except as fragments of a story too big for all of them. It is an ice-story, a cold tale, a quest riddled with cynical observations. The innocence found (or imagined) by the Thule expedition is long gone—or at least evolved in the cross-cultural confrontation that is today's reality.

Smilla herself is a character who has seen both worlds and knows the North as home:

I'm not perfect. I think more highly of snow and ice than love. It's easier for me to be interested in mathematics than to have affection for my fellow human beings. But I am anchored to something in life that is constant. You can call it a sense of orientation; you can call it woman's intuition; you can call it whatever you like. I'm standing on a foundation and have no farther to fall. . . . I always have a grip—with at least one finger at a time—on Absolute Space.[5]

The sense of snow is not a matter of terminology. Here is a character most comfortable with principle, a mystery that strives not toward resolution but after emptiness. There is something strange and terrible guarded and hidden in the far reaches of Greenland. At various times in the story it seems like it might be priceless, alive, alien, or technological. In the end it is none of those things. The North must retain the sense of mystery through the end, and as the conclusion of this now-bestselling thriller keeps us reading to the final page, we all must learn that the answer to the puzzle never supersedes the question. The plot echoes *Frankenstein* as the pursuit comes to be frozen in by ice. Running, gaping across the white wilderness in the final North. "It's only the things you don't understand that you can resolve."[6] This book transcends resolution to escape the pitfall of the mystery story, our failure to care once the plot has been thickened to taste. For in the North nothing can be as tangible as the palatability of white space. Don't think any story, however compelling, can put a dent in it.

We read these journeys, we wish to join them. We wish they could be possible today, now that wilderness is bound and gagged on so many maps. I imagine I have been with Rasmussen, listening to Nalungiaq. I have set her words to music.[7] The retrace of the journey is strong whenever one wants the North, hoping a current experience can gel with one from the past. Christoph Ransmayr's *Terrors of Ice and Darkness* recounts the 1981 quest of Joseph Mazzini, a man obsessed with the Austro-Hungarian North Pole Expedition of 1873. The modern and, of course, doomed adventure is laced with quotes from the articulate notebooks of Julius Payer, cartographer, painter, and first lieutenant of the original Imperial Expedition. Payer's musings depict exactly the appeal of surrogate isolation, the dream that we may reinhabit the emptiness of the past in the most expansive places in the present:

The road to the Arctic interior is a hard one. The wanderer who travels it must summon all his mental and physical energies to wring some scant knowledge from the mystery he hopes to penetrate. . . . He spends years in the most dreadful exile, far from friends, from all the pleasures of life, beset by dangers and the burden of loneliness. Only the ideal of his goal can support him; *otherwise he will wander, a victim of mental ambiguity, through an internal and external void.*[8]

Given this vision, what seeker would not want to go? The North becomes the perfect place for philosophy, the ideal location for the invention of the idea of wilderness. For it does not talk back, it only always sings with the wind.

I never thought it was so lonely. The people there twinkle with the abstract expanse. But then we do find what we look for. In openness we conjure our own places—the *Norths* are a plural place; there is no one metaphoric pole where the compass needle of our visions needs to point uniformly. The truth of experience is never so uniform.

The same space opens up to those arts that wish to fill it without words. Glenn Gould, the pianist, decided early on he would not play live for people, but live in his studio assembling versions of Bach and Mozart by splicing together a multitude of takes. The perfection of his music collages is legendary. Though once radical, his technique is almost standard in the world of classical music today. What remains eccentric is his extremity, his drive to escape and live in a world composed only of music.

What has his cutting and pasting to do with the North? "Living in the North," he hypothesized, spending most of his time indoors in Toronto, "makes one into a philosopher." He tried to articulate this feeling with a radio play, a collage of voices, stories fading in and out of life farther up toward the pole. I take my title from that of his work, "The Idea of North." Shall it be one direction, or many? I want to say "norths," pointing many ways. North, north, again somewhere else. *The* Norths, *Nordene.* Many places up and away. Gould gave up flying, and never got much farther than Churchill, on the shores of James Bay.

Canadian composer R. Murray Schafer, following in the splice-steps of Gould, has tried to explain what this philosophy of emptiness means for aesthetics: "The art of the North is the art of restraint. The art of the South is the art of excess." Schafer's music heads closer to the wild with each new year. He had a piece scored for dawn on a wilderness lake (see Chapter 10 in this book). Five thousand people showed up one morning outside Banff to hear it. Next he's writing for a week in the wilderness, an opera of participation only for those who are ready to give seven days of their time to submit to the work. The cold cannot be rushed. It is not for everyone.

Climbing the ridges up in Lofoten, mountains touching the ocean, unsung, too far North, unknown to climbers far away. Stettind. Svalbard. Torngat. Names not sure where they are. Their crispness is not in isolation, but in precision.

> *Start from the view:*
> *from blue half-islands that encode a sea,*
> *orange, perhaps, only cyan*
> *bands where*
> *sky equals cloud*
> *turn grayless gray*
> *wisped scree cliffs*
> *mass of water and air*
>
> *—just the Norths, nothing more—*
>
> *"You are now among those who seek out these places,*
> *go up to the empty Norths all around the globe,*
> *satisfied with latitude,*
> *habitations, ice flows, mountains of snow or men,*
> *none of it will stop the sky . . . "*

It is above a hundred degrees in lower Manhattan, one of those record-setting summer days. Laurie Anderson tells a story of hitchhiking to the North Pole in the heat. She stands out on Houston Street and flags down a passing car. "Hey—going north?" That's how it went all the way to Churchill, the end of the road at James Bay. From there it was a question of hailing planes, each leg a little bit closer to the goal. Sure it got cooler, but there's only so far one can go. The pole itself? Turned out to be a restricted area. No one was allowed to fly over unannounced. We all return back to the heat, after expending the limits of our desire to escape it.

These stories could go on and on. There will always be someone to reinvent the openness of northern climes and fill it with truth, lies, deceit, or imagination. Among the most curious is William T. Vollmann, a prolific novelist whose words pour forcefully out from the extremes of his own personal experience with the margins of society. Everything in his books seems to happen to him, yet it all sounds suspect, one shadow beyond the truth. In his latest novel, *The Rifles,* he appears to us in the form of Captain Subzero, a reincarnation of John Franklin, the legendary British explorer who died on his fourth trip trying to locate the impossible Northwest Passage. (The old folk song swirls down from memory: "Only the Eskimo in his skin canoe; was only one who ever got through.")

What sent Vollmann on his quest to dissolve fact into fiction? He himself asks: "Why did Franklin go north again? We who are interested in him mainly for his gruesome death believe that he did it to die, that he possessed a morbid lemming's heart whose ventricles were rimmed most dismally."[9] Vollmann overdoes it; he usually seems a little too strange for his own good, or his own fate. While recounting his own adventures in the ruined, drunk North of today's outer Canada, he (or Subzero or Franklin?) falls in love with an Inuit girl, gets her pregnant, and sets in motion horrible (or at least extreme) consequences. Did it really happen? Does anyone know? Hiding behind the mask of fiction, Vollmann eludes responsibility. His nickname in the margins is *Sallusautigiyattatuq*: "he never stops lying." This does not make the reader more comfortable.

Most interesting are Vollmann's detailed descriptions of what happens to sound when he is alone, trying to survive for a week in an abandoned weather station at forty below (Fahrenheit or Celsius) just to feel what Franklin felt as he lay dying: "The sound of something creepy coming down the hall was only the sound of his stove pressurizing. Likewise the sound of police sirens. The sound of bells was the sound of his frozen zippers clinking together when he walked."[10]

It's frightening, it's deadly, and it's a cold depiction of life alone, spared of the romanticism so believable in past journeys. Franklin is today a historical footnote, as much as he suffered, as driven as he was to forge on toward the usual explorer's doom. Vollmann, though, still wishes for something out of this emptiness to live for:

> There was a mountain straight ahead of him, soft, white and saddle-shaped. Its base was shaded in snow as in a fog. Above it, nary a cloud. He decided to walk toward it. He'd never have the chance again. The newly exposed strata of snow were hard and off-white like gypsum. There were lemming-tracks in the snow. He followed them. They paused in a wide and shallow hole in which the animal had left urine and a few specks of excrement like sunflower seeds, and then they went on and then a fox track joined them and then both tracks ended suddenly in the new-blown snow. As he walked toward the beautiful mountain, he was suddenly filled with pity for everything in the world, and he cried.[11]

So here is nature writing straight and unforced, full of fact and observation, with the crazy author putting himself on the line while still trying to see things for what they are. He lapses in and out of this directness. The fiction is impossible to separate from fact. The recounting of a journey turns happening into an object fixed and exemplary. There will be no chance to estimate its truth ever again.

The North now means more than it once did, newly opposed to the image of the South. What used to be a duality between the First and Third, the developed and undeveloped, Worlds, is now the polarity North versus South, supposedly emptied of directional prejudice. We of privilege, power, and energy abuse are now all the North, and those in poverty, struggling, oppressed by our greed and progress are all the South.

After all these spacious reveries, I retreat again into the story of another traveler. Tété-Michel Kpomassie recounts in his book, *An African in Greenland,* his childhood dream to reach that distant place where all is cold and white that is only hearsay in his tiny equatorial village. In his pilgrimage he seems to skip the polarity of the haves and have-nots in favor of a greater pull of the compass. He writes of the Eskimo who endows every object with life, who senses souls in all things. "In their eyes, all that drab, white, lifeless immensity of little intrinsic interest to an African like me becomes a living world." He learned to adjust, if not to blend in:

> I adapted so well to Greenland that I believed nothing could stop me from spending the rest of my days there. . . . But if I were to live out my life in the Arctic, what use would it be to my fellow countrymen, to my native land? Was it not my duty to return to my brothers in Africa and become a storyteller of this glacial land of midnight sun and endless night? Should I not open my own continent to fresh horizons, and the outside world? That is why I decided to leave.[12]

And we too all feel the pressure to return from the North, and from the wild. Anything can be carried to extremes, but what if left there? I cannot connect these with other stories. (I prefer words that leave space among themselves to those that try to tell us everything.) They float like hidden blocks of ice in the dark swirling sea. I have told you only the tips of them, those parts that rise above the surface into the sky of my memories. The world now has too many stories for us to weave any truth around them. In the North there is space, room to breathe, some would say nothing to do because the pressures of the manacled planet are so far from view. We remember only fragments, and these too recede quickly in the wake of the serious weight of the dark or the light. We can write down what we know, and tack it to the cabin walls. But soon these will freeze and be covered over with snow. We will run out of food. We will have to go home. Or learn to fish beneath the ice, to talk to seals and then kill them. And the distant troubles, far from our silent place, will also fall back and seem irrelevant.

Yet I know I'll return to the mess and the crime and the crowds and the neighbors who live too close together and smile or glance warily when they see each other in the baking streets. The North remains my opposite, even as part of me remains there and resents the oppression of the heat. Part of you is there too, even if you have never been. The antipodes of a circular journey, now I know that the North is not the wild. The wild is everywhere, always below the surface from the human point of view. It is North, it is South, East, and West. It surfaces when we accept that darkness and light make no sense without each other, when we see how each extreme can carry us away. But in most directions we miss what carries us, hidden beneath so much else we have made to cloak the original world. In the North these things are stripped away. We are left with the immediate, immutable nature whose depths remain frozen to the human touch. We might mine or flood it, but its brightness does not fall.

Carry us up to exacting spaces, open around from the earth to the sky. And leave us there not to die but to live.

Rest, Locate, Reckoning

run a finger along the top of the map
that's where I am

still here:

> *grabbing the frozen sand*
> *surprise and demand*

"we shouldn't go there"
"no one should go"

—that's what you learn once you've been—

no use complaining, others will follow

watch the ships come in, glowing in the nightsun
> *only outlines test the horizon*

how else will you know these things?
how else will I find you?

Notes

1. David Rothenberg, "Relative It Is," *Appalachia* (Spring 1988).
2. Gary Snyder, *The Practice of the Wild* (Berkeley: North Point Press, 1990), 73-74.
3. Christoph Ransmayr, *The Terrors of Ice and Darkness* (New York: Grove Weidenfeld, 1991), 1.
4. William Vollmann, *The Rifles* (New York: Viking, 1994), 15.
5. Peter Høeg, *Smilla's Sense of Snow* (New York: Farrar, Straus, Giroux, 1993), 45.
6. Ibid., 453.
7. David Rothenberg, *nobody could explain it,* (Somerville, Mass.: Accurate Records, 1991).
8. Julius Payer, quoted in Ransmayr, *Terrors of Ice,* 47. Emphasis mine.

9. Vollmann, *The Rifles*, 107.

10. Ibid., 274.

11. Ibid., 290.

12. Tété Michel Kpomassie, "Excerpts from *An African in Greenland,*" *Bostonia* (Summer 1993): 33-38.

Bibliography

Anderson, Laurie. *Stories from the Nerve Bible*. New York: HarperCollins, 1994.

Ferguson, John. "Northern Lights." In *Wild Culture: Ecology and Imagination*. Ed. Whitney Smith and Christopher Lowry. Toronto: Somerville House, 1992.

Gould, Glenn. *Glenn Gould Reader*. Ed. Tim Page. New York: Knopf, 1984.

———. *The Idea of North*. Radio Programme. Toronto: CBC Radio, 1967.

Hamsun, Knut. *Pan*. New York: Knopf, 1921.

Heaney, Seamus. *North*. New York: Farrar, Straus, Giroux, 1974.

Helprin, Mark. *Winter's Tale*. New York: Bantam, 1984.

Høeg, Peter. *Smilla's Sense of Snow*. New York: Farrar, Straus, Giroux, 1993.

Iyer, Pico. *Falling Off the Map*. New York: Knopf, 1993.

Kaysen, Susanna. *Far Afield*. New York: Vintage, 1988.

Kpomassie, Tété-Michel. *An African in Greenland*. New York: Knopf, 1981.

———. "Excerpts from *An African in Greenland.*" *Bostonia* (1993): 33-38.

Lopez, Barry. *Arctic Dreams*. New York: Bantam Books, 1987.

———. *Crow and Weasel*. San Francisco: North Point Press, 1990.

Millman, Lawrence. *Last Places*. Boston: Houghton Mifflin, 1991.

Moss, John. *Enduring Dreams: An Exploration of Arctic Landscape*. Toronto: Anansi, 1994.

Nelson, Richard. *The Island Within*. San Francisco: North Point Press, 1990.

Ransmayr, Christoph. *The Terror of Ice and Darkness*. New York: Grove Weidenfeld, 1991.

Rasmussen, Knud. *Across Arctic America: Narrative of the Vth Thule Expedition*. 1927. Rpt. Westport, Conn.: Greenwood Press, 1970.

Reed, Peter, and David Rothenberg, eds. *Wisdom in the Open Air: The Norwegian Roots of Deep Ecology*. Minneapolis: University of Minnesota Press, 1993.

Rothenberg, David. *nobody could explain it*. Somerville, Mass.: Accurate Records AC-4004, 1991.

———. "Relative It Is." *Appalachia* (Spring 1988).

Schafer, R. Murray. *Music in the Cold*. Toronto: Arcana Editions, 1977.

Snyder, Gary. *The Practice of the Wild*. Berkeley: North Point Press, 1990.

Vollmann, William. *The Ice Shirt*. New York: Viking, 1990.

———. *The Rifles*. New York: Viking, 1994.

10 . The Princess of the Stars: Music for a Wilderness Lake

R. Murray Schafer

CAST: Solo soprano, two mixed quartets or choruses, four actors, six dancers, and about ten canoeists.

ORCHESTRA: Flute, clarinet, trumpet, horn, trombone, tuba, percussion (4 players).

DURATION: 1 hour 20 minutes.

FIRST PERFORMANCE: Heart Lake, Ontario, September 16, 1982.

The Legend

"Without man the world was born and without him it will end"

This is the story of the Princess of the Stars,
daughter of the Sun-God and herself a Goddess.
Her name is in the stars and you have seen it there.
Each night she looked down on earth,
blessing it with kisses of light.
One night she heard a mournful cry coming up from the forest.
It was Wolf, howling at the moon, his double.
The Princess leaned over the forest to see who was singing,
but in leaning down so far she fell from heaven.
Suddenly she appeared before Wolf in a great flash of light.
But Wolf, frightened to see the stars so close.
lashed out at the Princess,
wounding her.

She ran bleeding into the forest, leaving dew wherever she went,
which was nearly everywhere, since she had no idea where to run.
By morning she found herself at the edge of a lake
and slipped into the water to bathe her wounds.
But there something caught her, dragging her down.
In vain she struggled:
in the end the waters closed over her.
You may see the stars of her crown at the tip of your paddle,
but the Princess you will not see.
The Three-Horned Enemy
holds her captive at the bottom of the lake,
and the dawn mist is the sign of her struggling.

The Action

All this has happened during the night. The story continues as the audience arrives at the lake in the darkness before dawn. In the distance we see a pinpoint of light moving slowly toward us from the opposite shore. At the same time the voice of the Princess is heard, singing an unaccompanied aria a kilometer away across the water. Her haunting aria has something of the quality of a loon. Soon other voices from far and near begin to echo it as the light-point continues to move toward us. When it reaches the shore, dawn has broken enough for us to see an old man in a canoe, the Presenter. He tells us of the happenings during the night, then turns and calls across the water.

Wolf comes to look for the Princess. He enlists the help of the Dawn Birds, dancers in canoes who arrive to comb the water with their wings; but they are prevented from rescuing her by the Three-Horned Enemy, who is keeping her captive beneath the lake. A battle develops but is interrupted by the arrival of the Sun Disc (sunrise), who demands to learn what has happened to the stars. The Sun Disc drives the Three-Horned Enemy away, sets tasks for Wolf before he can release the Princess, and exhorts the Dawn Birds to cover the lake with ice and sing there no longer until Wolf succeeds.

The Princess of the Stars is to be performed as a ritual in canoes at the center of the lake some distance from the audience, which is seated on the shore. The principal characters are either costumed and masked (as in the case of the Dawn Birds) or are enclosed in large, moveable structures fastened to the gunwales of oversized voyageur canoes depicting Wolf, the Three-Horned Enemy, and the Sun Disc. Since the characters in the canoes chant an unknown language, the audience is informed of the action by the Presenter, a kind of earth spirit or medicine man who acts as an interpreter between the observers and the performers.

The work is designed for performance at dawn on an autumn morning. It forms the prologue to my entire *Patria* cycle and introduces the central theme of the works to follow. From this lake, Wolf will go out in search of the Princess, seeking her forgiveness and compassion. If he can find her, he will also find himself. Then she will at last return to the heavens, and he, redeemed, will also rise to inherit the moon. Wolf's wanderings will take him to many distant lands and many historical periods before he returns to find the Princess in the same natural environment he deserted at the close of *The Princess of the Stars*.

The unifying motif of the *Patria* works is Wolf's journeys through the many labyrinths of life in search of the spiritual power that can both release and transfigure him. He will travel under many names and assume many guises, including the displaced human immigrant D.P., the Greek hero Theseus, a dead pharaoh seeking to be raised to heaven by the sun, or antimony in the "chymical marriage" or *hierosgamos* of the alchemists. At times he may assume great preeminence; at other times he may be chased away as a fool, a criminal, or a "beast." As the labyrinthine nature of his wanderings intensifies, the Princess becomes personified for him in the figure of Ariadne, who helped Theseus escape the Cretan labyrinth in the well-known Greek myth. The thread-gift provided by Ariadne in the *Patria* cycle is the thread of music. Ariadne's gift is her haunting voice; this is what sustains and transforms Theseus-Wolf during his life journeying.

Each of the *Patria* pieces is designed to exist on its own and many explore different theatrical settings and techniques, but all follow the theme of Wolf's search for his spirit in the guise of the Princess as it was introduced in *The Princess of the Stars.*

The Environment

The environment of *The Princess of the Stars* is extremely important, not only for the effect it has on the audience, but also for the ways in which it is intended to affect the performers. Its entire decor is what the Japanese call *shakei,* borrowed scenery. While the principal characters are on the surface of the water, the musicians and singers are positioned around the perimeter of the lake. The lake itself should be about half a kilometer wide and a kilometer long, show as few signs of civilization as possible, and have an irregular shoreline to allow the principal characters to canoe in from "off stage."

What distinguishes this from the traditional theatrical setting is that it is a living environment and therefore utterly changeable at any moment. The lighting alone is in a constant state of change, and atmospheric disturbances can descend or retreat to affect a good deal more than the audience's state of mind or comfort. Having witnessed a good number of performances of *Princess* I can say that it is a different theater every time. Will there be a sunrise? Will there be dew on the ground or mist on the water? What if it is raining or windy? Wind in particular is the enemy of the production, for it not only can make the canoes difficult or dangerous to navigate, but also can destroy the music, carrying the sounds off in unintended directions so that the audience loses much and the performers, who are without a conductor and who listen and react to one another by ear rather than eye, may miss vital cues.

The living environment thus enters and shapes the success or failure of *The Princess of the Stars* as much as or more than any human effort; knowledge of this must touch the performers, filling them with a kind of humility before the grander forces they encounter in the work's setting. But as we participate with these forces, allowing them to influence us in every way, is it not possible to believe that we as performers and audience are also influencing them as well? We disclaim belief that we can make the wind rest or the sun shine. Yet it is certainly not remote from the ancient peoples, whose rituals were often conducted in the open environment. Then one danced to make the rains come or cease or to make the corn grow or the caribou appear. There must be something of that kind of faith in the minds of the participants as they approach a performance of *Princess*. I have seen beautiful performances in the Rocky Mountains when the sun stroked snowcapped mountains just at the moment of the Sun Disc's entry, and I have seen haunting performances when the lake was enshrouded in thick mist. The production of this work will always be tinged with the excitement of a premiere. It will be a theater of first nights only.

Though the text of *The Princess of the Stars* is original, the work is clearly related to Indian legends, for like them it employs a story to account for various natural phenomena. There is dew on the grass because the Princess ran through the forest; the mist on the water is the sign of her struggle to be free of the Three-Horned Enemy. The Dawn Birds appear at precisely the time the real dawn birds are waking up, and singers and instrumentalists around the lake coax the real birds into song by imitating their calls. To know that one is affecting the environment in this way can fill one with awe. On more than one occasion loons or Canada geese have flown across the lake in front of the bird-dancers, mixing the real and the imitative to beautiful effect. The entire work is timed according to the sunrise; for the real sun should synchronize with the arrival of the Sun Disc, who has received messages of the distress on earth from the birds that awakened just before him. When, at the conclusion, the Presenter tells the Dawn Birds that the lake soon will be covered over with ice and snow, we know that this is precisely what will happen.

What is the effect of treating nature in this way? *By mythologizing the fluctuations of nature we have intensified our own experience of it.* We begin to flow with it rather than against it. We no longer spite it or shut it out as we do in covered theaters. This is our stage set, and we have become one with it, breathing it, feeling it in all its mystery and majesty. Of course, there will be problems, for nature is fickle (though, as Jung reminds us, she is never deceitful like man).

The Princess of the Stars is conceived for a situation. It should only be presented when the conditions are right. But since no producer can predict this and no union can enforce it, uncertainty is both its birthright and its stigma. Like the art of ancient times, it is wedded to its time and place by indissoluble links that guarantee that no counterfeit experience could ever replace it.

The Tempo

A production of *The Princess of the Stars* is never fast. Even when Wolf and the Three-Horned Enemy engage in battle, they move at the pace of an armada rather than that of participants in modern warfare. It takes the Presenter fifteen minutes to come down the lake to the audience area at the beginning of the show, an act that prepares us for the tempo of the entire work. But something strange happens when nothing happens. The senses are sharpened, becoming ready to print the decisive action when it occurs. There can be little doubt that primitive rituals are deliberately structured in this way. The punctuation of long, informationless interludes by sudden events causes an adrenalin rush to the brain, making the experience memorable.

And so in the slowness of the breaking dawn we are alert to the smallest change. Perhaps our eyes wander to the hills and notice they have become lighter. Or we notice a ripple on the water as a breeze skims across it. Or we hear an animal scurrying for cover in the underbrush. (On one occasion an instrumentalist had an animal run between his feet in the darkness.) In the slowness tiny events become magnified; large events are haunting. I have seen small children transfixed as Wolf comes onto the lake with lowered head swinging from side to side in the long hunt before he arches his neck and lets forth a blood-freezing howl.

The music of *Princess* also participates in this slowness. Since sound travels at slightly more than 310 meters per second, it takes the music of a performer at the far end of the lake three seconds to reach the audience and another three seconds for the echo (if there is one) to return. When we performed *Princess* at Two-Jack Lake in the Rocky Mountains, we gave the singers megaphones with which to focus the sound, bouncing it now off one mountain, now off another. The idea of employing megaphones in outdoor performances is certainly not new. There are diagrams of such instruments in the writings of Athanasius Kircher (*Phonurgia Nova*, Campidonae, 1671, pp. 117-43). And Sir Samuel Moreland in 1672 published a brochure, *Tuba Stenturo-Phonica,* describing a twenty-one-foot horn he made that could project the human voice up to a mile.

If the lake is surrounded by mountains and forests, echoes become a special feature of the performance. The feedback in the echo begins to modify the production, which lengthens and lingers in breadth and resonance as performers learn how to turn it into advantageous cybernation. At Two-Jack Lake we were able to position the musicians so that their sound was funneled indirectly to the audience area, the auditor sometimes receiving the echo more prominently than the original sound. Indirect reception removes much high-frequency sound, giving the tone a mysterious, remote quality—an effect, by the way, that was sought by Wagner, though in quite different circumstances.

The Ritual

In one sense the comparison with Wagner is not inconsequential. Wagner wanted to hide his musicians in a pit because their appearance distracted from the drama. The result was the creation of a *mystischer Abgrund* (mystical abyss) that separated the audience from the action on stage, thus taking on many of the aspects of a ritual. In *The Princess of the Stars* the musicians are hidden from view by the trees; we do not know from where the sound will issue next; and the action is distanced from the audience by the lake, for only the Presenter comes close to shore.

But there are other features that draw *Princess* closer to ritual. When he speaks to us for the first time, the Presenter tells us that before we may observe the "sacred actions" performed on the lake by "gods and animals," he must first prepare us. In a short thaumaturgical incantation, he turns us into trees in order that we may observe the events without interference. In one sense this distances us from the action; in another, it implicates us as part of the natural decor of the production. We are reminded of the motto of the work: "Without man the world was born, and without him it will end."

Language also contributes to the ritual effect of *The Princess of the Stars.* The gods and animals speak an unknown language—or to be more precise, a series of unknown languages—that the Presenter alone can interpret for us. Wolf chants an invented language incorporating some morphemic and phonetic elements of North American Indian dialects. This lends him an ancestral dignity, but has practical significance as well, since Indian languages have an abundance of long vowels, and contain few labials (such as *m*) or compact vowels (such as *i* [bit] or *ê* [bet]), which do not carry well in the open air. Similar considerations also affected the choice of instruments: for instance, tests proved that log drums, so dull and muted when played indoors, take on an exciting resonance in their natural environment, particularly if they are placed over pits or gulleys, which act as sound boxes. And so log drums and tom-toms became Wolf's accompaniment, four sets of them, played by drummers at the four corners of the lake.

The singers' texts contain actual Indian words for *star, lake, princess,* and *wolf.* These are not employed syntactically; they are color words chosen from a cross section of languages, predominantly those of the eastern woodlands. Another series of color words form the Sun Disc's welcoming music. Beginning with the ancient Japanese word for sun, *ohisama,* sun words follow in a geographic curve around the world through the languages of Asia, Africa, and Europe. The Sun Disc himself speaks an invented tongue with a strong suggestion of Latin cognates. The Three-Horned Enemy's speech contrasts sharply with that of Wolf and the Sun Disc; neologisms, notable for their monosyllabic abruptness of compact vowels and waspish consonants, are given additional bite and distortion by means of a loud hailer implanted in the Enemy's costume. A further independent vocabulary is that of the Dawn Birds' chorus, which derives partly from ornithologists' notebooks and partly from personal listening experience plus imagination.

Enough about the polyglottal nature of the work; it is an affair that will recur repeatedly throughout the *Patria* cycle, though not always with the same intent. The important thing to stress here is that when we come to witness a performance of *The Princess of the Stars* we are already removed from the action as suggested by the real and implied antiquity of the languages its characters speak.

All rituals are rooted in antiquity, or must appear to be. If they have not been repeated uninterruptedly throughout the ages, archaic dress, conduct, and speech can assist in creating this impression. When we performed *The Princess of the Stars* on Two-Jack Lake, gaunt black-robed ushers conducted the audience from the road to their places at the edge of the lake. In a more complex handling of ritual, more elaborate preparation ceremonies (including the consecration of the site) may be desirable, but here "holy nature" and the strange timing of the event seemed sufficient.

When the site is legendary its attraction is intensified. In his review of *Princess,* the late Banff poet Jon Whyte related a legend from Lake Minniwanka, which immediately adjoins the site of the production. In 1909, an Indian family was crossing the lake carelessly singing:

Suddenly out of the water appeared the huge back of a fish many yards broad, only to disappear, when out shot a beautifully shaped arm and hand, which clutched not in vain at one of the singers. Immediately a companion seized a knife and stabbed the arm through and through. The hand only clung the tighter to its victim and the surrounding waters were churned and lashed about as if the winds of heaven were let loose all at once.[1]

So there was a monster in the lake. And there was a woman possessed. Who can say she was not a princess? Of all the things said at performances of *The Princess of the Stars,* to me the most beautiful was that of a four-year-old child who murmured to her mother after all the characters had departed, "The animals were just pretend, weren't they, but the Princess was real."

We depart from the lake as the sun rises majestically above the mountain peaks, bathing them in crimson and gold. Is this the same sun that left us yesterday, or is it not the Lord of the cosmos who has set his commands for all to hear? Perhaps for just a moment we feel what D. H. Lawrence intended in *Apocalypse* when he wrote:

> Don't let us imagine we see the sun as the old civilizations saw it. All we see is a scientific little luminary, dwindling to a ball of blazing gas. In the centuries before Ezekiel and John, the sun was still a magnificent reality, men drew forth from him strength and splendour, and gave him back homage and lustre and thanks. But in us, the connection is broken, the responsive centres are dead. Our sun is a quite different thing from the cosmic sun of the ancients, so much more trivial. We may see what we call the sun, but we have lost Helios forever, and the great orb of the Chaldeans still more. We have lost the cosmos, by coming out of responsive connection with it, and this is our chief tragedy. What is our petty little love of nature—Nature!!—compared to the ancient magnificent living with the cosmos, and being honoured by the cosmos![2]

The Pilgrimage

For centuries our art has been produced for indoor environments. I will not dwell on how this has altered music, how it has conditioned the search for purity of expression and the suppression of all distractions, bringing about an intensification of acoustic image, high-fidelity, amplification, and sound presence. Everything has been arranged to enhance these values.

Musicians sit in a group on the opposite side of the building from the listeners, often disciplined by a conductor. They have set up unions to ensure that their working conditions are appropriate. Times of work and rest are dictated by contract. Light intensity must be sufficient; adequate warmth and controlled humidity for instruments must be guaranteed; proper sound amplification must be maintained. For this a whole battery of nonmusical specialists is engaged: stagehands, caretakers, lighting engineers, acousticians, recording engineers. The public needs to be informed of the undertaking, and to do this another fleet of experts is required: publicists, box office attendants, managers, promoters, printers, critics. All this apparatus is very costly to maintain. It has to be paid for one way or another, and if the public cannot or will not do so, foundations and government agencies will be expected to come to the rescue, bringing new confederates onto the scene: boards of directors, accountants, planners, arts administrators, more managers, heiresses, royalty, and so forth.

This model of the musical or musical/theatrical undertaking was once very efficient—in fact, such a paragon of efficiency that by the eighteenth century it actually served as the model for all future capitalistic enterprises. Laboring professionals were harnessed and yoked under the leash of a foreman, promoted by expectant owners and backers before the first paying public in history. This commodification of art extends today to all the auxiliary agencies that expect to make greater profits from it: publishers, broadcasters, recording companies, and performing rights societies. I do not allege that this model works efficiently all the time. In fact, as time goes on, it works less and less effectively—or at least only works effectively when the cash flowing in from the public exceeds the cash flowing out for the services.

The Princess of the Stars is a work that, simply by moving outdoors, challenges us to breathe again. But disguised beneath its simple plot and musical textures are timely questions—big ones. When musicians play across a lake at a distance of half a kilometer or more, how are they conducted or supervised? Of what value, then, are conductors and managers? When everything is muted at long range, like the tinting of colors at a distance, how is presence to be obtained in the sound? Or is this ideal to be abandoned? How do performers cope with unpredictable weather, cold, dampness, transporting themselves and their instruments over uneven terrain? Can we arrange a contract to remove the hazard of rain? And what about the managerial staff? We need boatmen, not caretakers; trailblazers, not electricians; naturalists, not publicists.

And the audience? Instead of a somnolent evening in upholstery, digesting dinner or contemplating the one to follow, the audience sees this work before breakfast. No intermission during which to crash out to the bar and guzzle, from which to slump back after a smoke. No pearls or slit skirts. It is an effort to get up in the dark, drive thirty miles or more to a damp and chilly embankment, then sit and wait for the ceremony to begin. And what ceremony! Dawn itself, the most neglected masterpiece in the modern world. To this we add a little adornment, trying all the time to move with the elements, aware that what can be done will be little enough in the face of it all. Here is a ceremony rather than a work of art. And like all true ceremonies, it cannot be adequately transported elsewhere. You can't poke it into a television screen and spin it around the world with anything like a quarter of a hope that something valuable might be achieved. You must feel it, let it take hold of you by all its means—only some of which have been humanly arranged. You must go there, go to the site, for it will not come to you. You must go there like a pilgrim on a deliberate journey in search of a unique experience that cannot be obtained by money or all the conveniences of modern civilization. Pilgrimage is an old idea, but when more than 5,000 people travel to a remote lake in the Rocky Mountains to see a performance of *The Princess of the Stars,* it is evident that this old idea is one for which there is a contemporary longing.

Notes

1. Jon Whyte, *Crag and Canyon,* Banff, Alberta, August 14-20, 1985.
2. D. H. Lawrence, *Apocalypse* (Harmondsworth: Penguin, 1974), 27.

IV . THE WILD REVISED

11 . Beauty and the Beasts: Predators in the Sangre de Cristo Mountains

Tom Wolf

Man has won. The wilderness killers have lost. They have written
their own death warrants in killing, torture, and blood lust, with
almost fiendish cruelty.

Arthur Carhart
The Last Stand of the Pack (1929)

Pity would be no more,
If we did not make someone poor.
William Blake

The killing fields have shifted to our cities. As surely as Colorado's urban violence escalates, especially in my old Denver neighborhood of Park Hill, our
rural violence declines. Having eliminated our competitors, we prefer to
prey on each other.

That's what gives predators like the wolf, extinct in Colorado for more
than fifty years, another chance, maybe more than just another tragic last
stand. Colorado's remote Sangre de Cristo Mountains, oddly enough, present us with the chance of a lifetime—the chance to reinvent our relationship
with wild animals, our fellow predators—the chance for mutual respect.[1]

Hope doesn't have to wait for the government to act. Private individuals
working on private land already have begun the miracle of raising the Sangres from the dead. I learned this hope for reclamation of the Sangres while
breakfasting recently at the Sands Restaurant in Alamosa, Colorado. While
the *Denver Post* regaled me with the latest drive-by shootings in Park Hill,
the morning regulars drifted in and took up their accustomed places around
a big table within earshot. They kick-started their morning with coffee and
tales of urban violence, accelerated through fifty different reasons to cuss the
feds, and finally settled into cruising speed on a topic of long-standing debate: the differences between dogs and wolves. A reader from boyhood of
classic wolfers such as Arthur Carhart and Aldo Leopold, I thought I knew
all about wolves. But those cruise-controlled caffeiners grasped the shape of
a wolf's paw, the intricacies of its jaw, the length and strength of its legs.

The people in the communities around the Sangres still live close to animals—close enough to travel the conceptual distance needed to work together with wild animals. Too often television and movies rend and render nature into hopeless sentimentality, something to be pitied. And whether as cause or effect, the media depict urban life as hopelessly violent, anything but urbane. Now my neighbors in remote places like the Sangres have the chance to heft the burden of our responsibility toward our fellow predators. Animal trainers know the perils and rewards of this responsibility. They learn the all-important lesson of humility, because they know that training means mutual mastery of a common language, a code by which an animal can tell you more than you may want to know about what it perceives when perceiving you. To see ourselves as others see us remains a worthy goal. In the case at hand, we can respect the wolf's "otherness" while exploring common ground: the Sangre de Cristo Mountains.

In spite of their blood lust against predators, we should honor conservation pioneers such as Carhart and Aldo Leopold. We should honor them all the more by following the spirit rather than the letter of their example: *question authority*—and this includes questioning their own authority. On August 13, 1993, President Clinton signed the Colorado Wilderness Bill, officially designating the Sangres as wilderness. Carhart and Leopold endowed us with this concept of wilderness, to be sure, but it is a wilderness without major predators, a wilderness without biodiversity, something like religion without God, or Self without Other.

They willed us this zoolike wilderness, but then they added the impoverished concept of wilderness carrying capacity. In Carhart's case, that carrying capacity tended to be cultural and aesthetic. His colleagues called him "the beauty doctor." In Leopold's case, wilderness carrying capacity leaned toward the biological. Leopold's ecology (which is today's official ecology) offers productivity and efficiency at the price of diversity. Together, these two giants impart to us the newly ordained Sangre de Cristo Wilderness as cathedral: beauty on a grand scale, overflowing with freeloading worshippers, but an empty tabernacle—and, perhaps worse, an empty collection basket.

It's unavoidable. This train of thought chugs through a West clear-cut of its heroes. Perhaps that is just as well. We can live without heroes, but we cannot live without hope. While there is little hope for biodiversity in highly visible "crown jewels" such as Rocky Mountain National Park, there is hope aplenty in the remote Sangres, even in the National Park Service (NPS)'s Great Sand Dunes National Monument, because in the Sangres the potential for cultural and biological diversity is still high. Amidst the obscure Sangres, people value personal freedom, individuality, and private property just as much as they value voluntary associations and public land. If wolves ever return to Colorado, it will be thanks to such determined individuals—not to federal agencies such as the U.S. Fish and Wildlife Service (FWS), which is in charge of recovering endangered species in their former habitats. After the debacle of its efforts to raise wolves in Rocky Mountain National Park, the FWS drew back. Way back. Until 1992, the FWS consistently refused to include any part of Colorado in Endangered Species Act wolf recovery zones. Inscribing its argument in a perfect circle, the FWS said it could not study potential wolf habitat in Colorado because there were no wolf recovery zones there. Colorado's congressional delegation finally had to force the FWS to begin the study.

Why Big Fierce Animals Are Rare

Big, fierce animals are rare because we made them so, because we continue to make them so. They remain rare because of perverse laws such as the Endangered Species Act, because of the implied state ownership of wildlife, and because of perverse institutions such as the U.S. Forest Service, the National Park Service, and the Colorado Division of Wildlife.

Don't allow the sensational cases of the wolf or the grizzly bear to sidetrack the argument. Though there certainly were grizzlies in the Sangres until the turn of the century, there were never very many. My life will end before they return. Since lines must be drawn, I draw the line at the grizzly, which may have a fighting chance across the San Luis Valley in the massive fortress of the San Juan Mountains. I accept the more adaptable black bear in place of the grizzly, since the bears' roles in the Sangres' ecology is similar enough to avoid the kind of unraveling that occurs in the absence of major predators.

Consider Lanky. I did. Though aware of each other, thanks to tracks and scat, Lanky and I did not lock eyes until one recent predawn, when a coyote chorus and a conjunction of Venus and Jupiter rousted me out early. Foolish and unwary with sleep, I stumbled in the near dark down to creekside, where I could just discern a pair of coyotes, dancing dervishlike over a dead doe in full bear hug. A gore-splattered, protective, 175-pound black bear boar will stand still only so long. Lanky growled me right straight off his kill. But before he returned to work, he showed me how a respectable bear fattens for the coming winter, simultaneously shitting and supping: predator.

At noon, I ventured back to the kill, which Lanky had buried (pending a nocturnal return) under scat, leaves, branches, and dirt. Only the stomach and its contents remained intact, clearly showing that Lanky owed his venison to his fellow predator. Someone had gut-shot that doe. She died before she could reach water.

Lanky hibernates now, as I write. Maybe that doe increased his chances of making it through a tough winter. My neighbor, a noted wildlife photographer, has "shot" many a bear. She christened him "Lanky" because her shrewd eye judged him too lean to winter well. Lanky is still a growing boy. Unlike the furtive, timid, stunted black bears of Rocky Mountain National Park, Lanky occupies prime bear habitat, and his bared fangs showed me he means to defend his newly occupied territory, while his Rocky Mountain National Park counterparts eke out an existence in the most marginal fringes of their home, an International Biosphere Preserve, leaving optimal areas to hordes of humans.

A few miles north of here at Cotopaxi, along the Arkansas River, a black bear killed and ate a man last summer after the man baited the bear with garbage and then wounded him with a poorly placed rifle shot. In the due course of such things, the Division of Wildlife shot the bear. Since the bear is the property of the state, the state acts quickly to avoid liability problems. What fate awaits Lanky? Because he has not yet been fully conditioned to fear and avoid humans, Lanky exemplifies a hard lesson: all too often, the more layers of "protection" we slop on a Rocky Mountain landscape, the faster its biodiversity declines.

Lanky and his human neighbors share the ponderosa pine/Gambel oak zone, living at the edge of the San Isabel National Forest on former marginal ranch land that was subdivided twenty-five years ago. Prehistoric and historic records show that Lanky is hardly the first bear to set up shop in the neighborhood, but is reoccupying former bear territory. Because we suppress fires and fence out cows, we raise a lot of wind-pollinated scrub oak. When not lush with the fragrance of venison, Lanky's scat reeks with tannin from acorns. When the inevitable fire next time consumes this vegetation association, we will all retreat. Studies of charcoal lenses in the soil and fire rings in the trees show, after all, that prehistoric fires burned in this vegetation type every five to twelve years.

Stability? Old growth? Not here. At least not until you ascend to harsh, high-altitude sites, where you may encounter some old-growth spruce in the 400-year range. There you may also find the true treasure of the Sangres, 1,500-year-old ancient ones: bristlecone pine and limber pine—the crooked timber of the Sangres. And yet, as numerous studies of Robert Peet and colleagues make clear, the forests of the Sangres are the most diverse (both in species and in communities) of any forests in the Southern Rocky Mountains, an area stretching from Montana to New Mexico.[2] It's just that there are a few important pieces missing in the predator category, pieces whose absence threatens to unravel the whole.[3]

If only we knew where "halfway" was, we might meet predators like Lanky there, in that landscape in the Sangres, and in that landscape of the mind where we both fear and respect each other. It's not like there is any shortage of prey for any kind of predator. Thanks to the Colorado Division of Wildlife's all-too-successful game management programs, today's Sangres overflow with elk and deer in such numbers that human hunters cannot keep pace with the increase. Meadows in the Sangre de Cristo Wilderness that sheep, horses, and cattle overgrazed for fifty years now experience a conifer invasion as they groan under heavy ungulate loads. If managers are not careful, biodiversity in the Sangres will soon suffer from the dreaded *too many*, the same lethal strain of hoof-and-mouth disease that afflicts our national parks. The potato-chip-addicted deer at browsed-out Great Sand Dunes are only the beginning. What would Leopold and Carhart say?[4]

During much of the year, to the embarrassment of some wildlife biologists, the feeding habits of elk and cattle are indistinguishable. Red meat is red meat. When we see riparian zones beat out by cattle on Forest Service land, we call it overgrazing. When we see the same phenomenon involving elk or deer on Park Service land, we call it "natural" management, an excuse for doing nothing. Yet lack of management may result in the steep declines in biodiversity that curse Rocky Mountain National Park, leaving its meadows carpeted with elk scat, its riparian areas denuded of beaver-sustaining and watershed-stabilizing willows and aspen.

Elk were rare in the Sangres by 1878 and locally extinct by 1900. In 1880, with gold and silver booming, there were 8,000 people in Custer County, on the east side of the Sangres. That population bottomed out in 1980 at around 1,000. It is increasing rapidly now. In 1888, Cockerell had to travel far to the west to hunt elk. The current population started from Wyoming stock railroaded into the Sangres after World War I, thanks to cooperation between the Forest Service and local volunteer service associations such as the now-defunct Sangre de Cristo Sportsmen's Association. That population expanded again until World War II, when poachers shot out what did not starve in competition with vastly increased cattle herds. In 1947 there were no elk in the Sangres and only thirteen in the roughly contiguous Wet Mountains to the east.[5]

Now no sheep, only a few horses, and only a few hundred cattle graze in the Sangres for any time at all. A year-round population of nonmigratory elk in the Sangres numbers in the tens of thousands, nicely demonstrating that managers can meet cultural and biological carrying capacities while ignoring biological diversity. Some of these elk so closely resemble cattle in their behavior that they spend all their lives on former grazing allotments and ranch lands, where forests are reoccupying the territory, and where subdivisions are proliferating. Though heavily hunted, these animals thrive because they find thermal and visual cover in the dense woods and rugged country that characterize the Sangres after a century of fire suppression. They do not even bother to move up and down the Sangres' elevation gradient with the seasons.[6]

Is this a success? Will the population collapse when range quality finally declines? Is this "success" also due to the rarity of predators? That's what Aldo Leopold thought. Yet even Leopold's famous Kaibab Plateau deer case study now seems deeply flawed. The traditional explanations for the population eruption were predator control and protection of does from hunting, while the decline was thought to have resulted from loss of food supply and habitat caused by overpopulation. But a closer look demonstrates that prior grazing by domestic livestock, fire control, and drought were much more important than Leopold's culprits.

In disturbance ecosystems like the Sangres (or the Kaibab), ungulate populations are not self-regulating. The Sangres owe their current "success" in red meat production to drought (producing a long series of easy winters), to nearly fifty years of relative freedom from domestic livestock grazing, and to aggressive fire-control policies made possible by large expenditures of federal funds.

This "success" is also due in part to the cooperation of private landowners, who can qualify for wildlife damage payments if their haystacks and alfalfa fields suffer. Since everyone benefits from such programs, including thriving populations of bears and lions, it is hard to see why such programs should not include the propagation of the wolf.

Maybe a quarter century ago, it was rare to encounter bears or lions in the Sangres. Now I often pause and reflect at the sight of lion tracks daintily set on top of mine when I find myself retracing a previous evening's walk. Does this mean that the Sangres will soon be unsafe for humans? Hardly, for unlike those of ungulates, the populations of big, fierce animals can be self-regulating. They don't necessarily need our help. They have their own ways of making themselves scarce. Only the zeal and the limitless funds of the federal government would push predator "control" to extinction. Only the bear and the lion could escape the fate of the wolf, and then only because the Sangres are so rough, and because these animals' official classification shifted from varmint to game animal. Until 1965, Colorado had a fifty dollar bounty on mountain lions. Then the definition of their value changed. They were worth hunting. In 1992, however, Coloradoans in their wisdom went to the polls and supported a referendum halting spring bear hunts. Now, in the summer homes around the Sangres, wildlife officers receive an average of one complaint per day: "problem bears" are doing so well on elk and deer that they are expanding and running into an expanding population of people.

Like wolves, mountain lions are big-bodied and therefore relatively rare in the Sangres. But their solitary social life differs from that of the gregarious wolf (parallel traits differentiate our relationships with domestic cats and dogs, as any trainer knows). And the lions, like the black bears, relish the rough-cut canyon country of the Sangres, making their extinction much less likely than that of wolves, even with the help of hunting dogs. Elk and wolves disappeared from the Sangres for the same reasons the bison and Utes did. On a seasonal basis, they had to occupy lower elevation, exposed prairie habitats where their social lives made them vulnerable to the ultimate predator, a rifle-armed horseman.

Because of its persistence, the lion earns a closer look. In general, population size simply reflects the position of a species in the food chain, where less energy is available at the top than at the bottom. Yet lions are also rare because they require large, contiguous habitats—at least 30 to 250 square miles of winter habitat.[7] A male's range overlaps that of several females, but he will not tolerate other males of his own kind.

Genetic studies show that the minimum viable population (MVP), the number of animals needed not only to escape extinction but also to retain enough genetic variation to adapt to changes over time, is fifty breeding lions. Twenty of these must be male, according to the lions' social dynamic. A population of lions must have at least 500 square miles to sustain itself over long periods of time. Of course, the lion MVP is also subject to accidental fluctuations in birth and death rates as a function of catastrophes such as disease or being killed by vehicles.

At roughly 350,000 acres, the Sangres considered as an island barely nudge the 500-square-mile figure. Linked to other areas through corridors such as the Promontory Divide that leads elk and deer from the Upper

Huerfano Valley to the Wet Mountains, however, the long, skinny Sangres begin to make some ecological sense. I never thought so until ten years ago, when my headlights began to surprise increasing numbers of elk, deer, and even pronghorn crossing the highways, where golden eagles were joining ravens and magpies at the Roadkill Café.

When Theodore Cockerell arrived in the Sangres in 1888, he was more concerned with plants and insects than with predators, partly because by the late 1880s most of the predators were already shot out. When it was almost too late, biologists learned that big predators, unlike insects, are extinction-prone. Reasons include a dispersed food supply, high metabolic demand, low metabolic rates, and the limitless subsidies that even today drive federal programs such as wilderness recreation, public lands grazing, and predator control.

Yet it is becoming increasingly clear that without big predators, ecosystems like that of the Sangres simply unravel in ways that make restoration unthinkable. War panic and political deals drove managers of the Sangres to grossly overstock their ranges and to tolerate the poaching of deer and elk during the world wars. When the inevitable effects hit the few remaining wolves after World War I, the predators simply turned to livestock, offering government hunters the excuse to exterminate them.

While wolves persisted in the Sangres until after World War I, they had disappeared from Rocky Mountain National Park by 1900. After the passage of the Endangered Species Act in the mid-1960s, wolves were supposed to return to Rocky Mountain in 1975. Nearly two decades later, nothing has happened—except for an explosion in the park's elk population that is severely stressing ecosystems already suffering from the fond attentions of too many people.

Before it adopted today's policy of natural regulation, the NPS regularly shot elk in the 1940s, and there was a drop in stress on range. But with natural regulation, the brainchild of Aldo Leopold's son Starker, elk reproduction rates did not drop. Nutritional stress due to overgrazed range did not adversely affect elk sex life. Now the same processes are beginning in the Sangres, especially with their designation as wilderness.

Official and Unofficial Predator Control Programs

Arthur Carhart was an ecological Jekyll and Hyde. So was Aldo Leopold. These two patron saints of wilderness became the bane of the wild as represented by predators like the wolf. Whether in the Forest Service or out, neither could see the difference between the good of the agency and the good of the lands entrusted to the agency. In the Sangres, the gap between these two goods yawns wider and wider after "wilderness" designation has been applied.

Both Carhart and Leopold understood the threat the predatory National Park Service presented to the Forest Service. Good intentions to the contrary notwithstanding, Forest Service administrative wilderness designations competed (for congressional favor) with the mushrooming offerings of its rival. The Park Service insisted on creating and then satisfying a public demand for highly managed, sanitized experiences that contrasted sharply on paper with Carhart's proposal to make Trappers Lake off limits to housing developments. Carhart made Trappers Lake safe for fishermen—too many fishermen. Neither agency made any place for wolves.

Though always present, like some low-grade infection, Park Service-Forest Service competition has never hit the Sangres very hard, since Great Sand Dunes National Monument was created from lands the Forest Service (at the time) did not want. In fact, local forest rangers were among the primary proponents of the monument, now one of the most heavily visited in the Southwest. Today, we can see that maintaining biodiversity in the Sangres becomes nonsense without including both the Dunes and their backcountry, the Medano Creek watershed, which is home to the endangered Rio Grande cutthroat trout and potentially home to more native fishes of the Upper Rio Grande Basin.

Official killing of predators was supposed to cease in 1936, when the parks were supposed to become, in some sense, sanctuaries. But for whom and for what? The enabling legislation for Great Sand Dunes National Monument addresses neither of these subjects.

Consider, for instance, the case of Medano Creek. The monument's enabling legislation refers to Medano Creek, whose flow arrests the fugitive, 700-foot-tall dunes, washing them, grain by grain, back toward the Rio Grande. Somehow the Park Service gained a federally reserved water right to Medano Creek. That was the key to defeating a water export proposal that might have destabilized the dunes. Whether this victory is hollow depends on what the winners do with their momentum. Can and will the NPS take seriously its newly understood role in maintaining the biodiversity of the Sangres?

New plans for a bigger parking lot and a new visitors center at Great Sand Dunes do not bode well for the future, since these facilities will stimulate off-road-vehicle use of Medano Pass. While the Park Service charges its usual nominal fee for entrance, the Forest Service charges nothing at all, in spite of obvious and expensive impacts to the watershed. It is hard to see how bisection helps the Sangres. Since the road follows Medano Creek through the monument and onto Forest Service land, increased traffic may harm endangered species recovery programs. To compete for recreation, both the Forest Service and the Park Service have introduced game fish, such as the rainbow trout, into this watershed, creating competition for the Rio Grande cutthroat trout. Furthermore, traffic on the Medano Pass road must travel east through the huge Wolf Springs Ranch, probably the best potential wolf habitat in the Sangres. No one has explained to the owner of this ranch how and why he should put up with this traffic's impacts on his ranching-for-wildlife operation. How do we get ourselves (and the Sangres) into such problems?

Leopold and Carhart saw that building a political constituency for wilderness and for wildlife management in Congress and in state legislatures meant the cultivation of a fishing and hunting constituency that not only had no use for wolves but saw them as dangerous competitors for scarce big game. Handing the new profession of wildlife management to the states and to state-supported land grant universities was a stroke of genius. Big game could be managed best by the states, and the insane job of killing all wolves was best entrusted to the latest in a long line of renamed and relatively obscure federal agencies, the Bureau of Biological Services (BBS), where no one would ask too many questions about the marginal cost efficiency of systematically destroying every last wolf in the continental United States outside Alaska.

Not to be outdone, the National Park Service (and the Audubon Society) actively conspired in this project, even though formally the parks were supposed to be wildlife sanctuaries.[8] A peculiar legacy of this collusion came to light as the Park Service and the Forest Service chose to monumentalize Great Sand Dunes National Monument in 1984 by renaming a nearby 13,297 foot peak after local legend Ulysses "Ulus" Herard. Herard did more than anyone but Stanley Young to decimate predator populations in the Sangres and make the Sangres safe for the Park Service and for the domestic stock that the Forest Service was pushing higher and higher into the mountains through an aggressive road and stock driveway building program.

That the Sangres are a good place to grow predators can be seen from the story of Herard, who ranched on Medano Pass until 1940. Herard's father had first seen Medano Canyon in 1849 on his way to the California goldfields. He later returned with his family to the Sangres, where he positioned himself in the center of the best predator habitat in the entire range. By 1890, far in advance of the Park Service, the Herards were running a rainbow trout hatchery at their ranch on Medano Creek.

In 1878, a mule kicked young Ulus Herard in the head, leaving him stone deaf for life. Ulus's disposition (and his attitude toward government) did not improve when a forest ranger ran off with his wife. Local legends abound detailing Ulus's revenge. Taking advantage of Forest Service range access help, but completely ignoring their attempts to regulate his activities, Ulus until 1935 ran 1,000 horses and anywhere between 1,500 and 6,000 cattle on a huge realm of around 125,000 acres that stretched east from the dunes, over Medano Pass, and down to Redwing and Gardner.[9] Ulus did all this alone; the stock ran loose over the commons, and he took what he needed at roundup time.

What this means is that the country was productive enough to support such numbers, at least temporarily, although Forest Service files overflow with reports about the adverse consequences for rangelands and watersheds. Meanwhile, especially during the world wars, the Forest Service was cultivating its livestock allies by building stock driveways at public expense into every part of the Sangres. Predators of all kinds attempted to reap this whirlwind, as evidenced by the many stories of Ulus's prowess in slaughtering his competitors in great numbers. Ulus was one of the few ranchers in the Sangres who did not ask for or need BBS help. He claimed 100 lion kills, as many black bears, and even a few grizzly. Anyone who considers the Sangres a poor candidate for predator propagation and re-introduction should study Herard's story—not to cluck disapproval, but to see what potential is there for the future. A Mount Herard overlooking a healthy wolf population would be well-named indeed!

While the Denver-based Carhart was drawing up his vast recreational plans for the Sangres, Herard's tale exemplifies what was really going on in the Sangres with respect to biodiversity, to predators. Not surprisingly, elk populations plummeted during these times, reaching all-time lows in 1936, when, for example, the Forest Service reported no elk at all in the Upper Huerfano drainage.[10] How did this split between recreation and diversity develop? How did Beauty lose contact with Beast?

Carhart in the Sangres

Born in 1892 in Iowa, Arthur Carhart brought a midwestern messianism to conservation. As the Forest Service's premier recreation engineer (his actual title), he had to move quickly to make his mark when he was hired in 1919. Sensing an opening, the resourceful Carhart immediately descended upon the otherwise worthless San Isabel National Forest with the kind of staff officer bravado that makes common-line people cringe, grimace, and roll their eyes. In the Forest Service's quasi-military line of authority, Carhart ordinarily would have made himself as welcome as wasps. But the San Isabel had been blessed with an extraordinary forest supervisor, Al Hamel, who in 1919 discerned unsung values in the forgotten Sangres. Flamboyant Carhart went on to change Colorado, if not the Forest Service or the San Isabel, in profound ways. He created our idea of the beautiful—and the safe (i.e., predator-free) wilderness. And he prepared the way for the second tragedy of the commons that the Sangres face today.

The year was 1919. In the entire history of the young Forest Service, Congress had never made an appropriation for recreation, and recreation was what the Sangres seemed to offer the ambitious. More or less contrived and fraudulent wartime beef and wool shortage scares had both forced and allowed Hamel to overstock the Sangres. As depleted ranges yielded dwindling revenues, Hamel went fishing for other sources. He caught Carhart.[11]

In the spring of 1919, before Carhart ever laid eyes on still-snowed-in Trappers Lake, the two men toured the San Isabel by automobile—an important breakthrough. Like so many conservationists of their time, Carhart and Hamel perceived the national forests as levelers, as democratizers, as extensions of the Jeffersonian concept that every American should own workable land, whether the crop be potatoes or patriotism. Taking his cue from Gifford Pinchot, Carhart bitterly attacked those who would set aside parts of the forests for private use or special interests. (He did not oppose private interests as such—in fact, he organized and encouraged them—but he did oppose private greed at public expense.) Yet we must still judge Carhart harshly today. Unfortunately, he did not see that BBS predator extermination programs benefited only special interests in the livestock, hunting, and recreation industries. All of these found ways of building special subsidies for their narrow interests into agency budgets. None of them worked for the value that I hold highest: biodiversity in the Sangres.

After he and Hamel parted, Carhart presented his plans to Carl Stahl, then regional (at that time, called district) forester. Ever alert to land-grabbing, budget-busting threats from the nascent National Park Service, Stahl had to endure the presence of Rocky Mountain National Park in the midst of national forests under his administration. Trail Ridge, the alpine highway bisecting the park, must have seemed a brilliant and threatening move in that high stakes game. The Park Service had already cut its deals with the railroads, which were only too glad to run lines to major terminals at the showier parks. How could Colorado's dispersed and difficult-to-access national forests ever compete? Which agency would control the future of automobile-powered mass recreation?

Stahl wisely sent Carhart to New Mexico to meet with Aldo Leopold, then the assistant district forester in Albuquerque. They met on December 6, 1919, at a time when Leopold's energies were hell-bent on the eradication of predators, not the edification of tourists. Leopold saw sportsmen as the political allies of the Forest Service's future. Carhart presented his older colleague with a bold new bridge between their obsessions. The two men agreed that wolves and other major predators had to go, but they also agreed that some Forest Service lands should be maintained in what they saw as a "natural" state. Somehow, the flash of friendship between these freethinkers begat wilderness.

For all his long life, Carhart would find ways to push the Forest Service toward plans big and bold enough to include beauty and (nonpredatory) beasts. In the 1930s, as civilian administrator of federal Pittman-Robertson funds, he would return beaver to the trapped-out, grazed-out drainages of the Sangres. But earlier on, he had a lot to learn. In 1922, Carhart the utopian dreamer submitted a plan to the sober Stahl. He proposed expanding the Forest Service's recreation budget to $56,000 and hiring five fellow landscape architects. Instantly red flags flew. Opposition to his plans came not only from grazing, mining, and timber interests, but from the Park Service's bulldog of a director, Steven Mather. In a memo to Carhart, Mather said that only the parks were set aside for recreation; National Forest lands were to be used strictly for watershed protection and commercial development. Fearing both bark and bite, Stahl was not the first forester to back down at the snap of Mather's jaws.

With this interagency whipping stinging his backside, Carhart then had to deal with a congressional mauling—a stingy recreation appropriation of $900. Disgusted with the lack of congressional support and Forest Service political compromises, Carhart quit in 1922, even though Stahl approved his Trappers Lake plans that same year. Later he would learn to manipulate both Congress and the Forest Service from outside vantage points. He would learn to follow money.

For Coloradoans, Carhart's brief career with the Forest Service was at least as influential as Leopold's long one. Far from a failure, he showed how later generations might push the Forest Service into big, bold recreational planning of a sort that now comes back to haunt the Sangres as they enter their designated wilderness era. And yet Carhart's legacy also includes other ironies and ambiguities. Today's Forest Service still strives to keep the Sangres safe from predators. By filling the Sangres to the brim of their cultural and biological carrying capacity, Carhart's and Leopold's heirs jeopardize the Sangres' biodiversity, their true (though still secret) claim to fame.

Arthur Carhart never relaxed his grip on the Sangres, on predators, or on the Forest Service. In an avalanche of 5,000 publications, he hammered away at the same theme: the right of every American to the primal pioneer experience, to coming into the country, to owning a piece of the West.[12] Carhart never would change the tune he established in his 1920 *San Isabel National Forest Recreation Plan,* an immodest attempt to pump his superiors for money. That plan "forecasts a time when in all of the Forests of the Nation a really comprehensive plan for regional development will be in force and, by a correlation of the recreational use with other activities and a full utilization of that use consistent with the best use of the Forest, will give to the people of the Nation the fullest return possible from their Forests. The San Isabel Plan is a pioneer plan of the type, but it is so well founded on common sense need and rational utilization of possibilities that it will probably stand for a long time as the model of big recreational planning in our Forests. This plan is truly an answer to a recognized need and a step towards full Forest utilization."

In pursuing their internal propaganda blitz, Carhart and Hamel had already done their local spadework with the San Isabel (Public) Recreation Association (SIPRA). Carhart pointed out that the local tourist industry was willing to put up big money to support recreation developments on public lands. He knew no bounds in his zeal to woo internal and external support. Perhaps in their hurried tour of the San Isabel, he and Hamel bypassed Ulus Herard's private predator extermination project on Medano Pass. Perhaps they also sped past Colorado Fuel & Iron (CF&I)'s Orient Mine. The history of southern Colorado's labor problems, culminating in the 1913 Ludlow Massacre, intimately involves CF&I, SIPRA's biggest financial supporter.

But Carhart the utopian dreamer had a solution to most regional problems: recreation. He wanted to develop a "system of camps for the industrial population of this portion of the state. The thousands of citizens of foreign birth or of foreign extraction found residing near the borders of the Forest and now never realizing anything from this proximity, will through cooperation of the Forest and with the Industrial Companies, come to know the hills and by means of camps where it will be possible to live as cheaply or even at a more reasonable rate than in town, these people will become better citizens and far less open to insidious suggestions of the radical agitator to strike at this land they have come to know and love."

Ever upbeat, Carhart labeled the forest "the San Isabel Playground." In those distant times before air conditioning, when my Kansas, Colorado Springs, and even Denver relatives retreated to the mountains, he called attention to the heat of the summer, especially in nearby Texas and Oklahoma. Like the builders of Trail Ridge Road at Rocky Mountain National Park, he had big plans for developing the former toll roads over Hayden and Mosca Passes, connecting the latter with Medano Pass and with Pass Creek Pass— all as year-round roads. A believer in the teeming masses-moving power of railroads, Carhart (like so many Americans of his time) also saw the potential for the automobile, which pioneered beyond the rail terminals to access a "system of uniformly built, uniformly managed hotels, working together."

Having identified predators as Public Enemy Number One, he wasted no time in fingering Number Two: "It is probable too that there will be an opportunity for a good deal of active propaganda against the fire menace in the camps and picnic spots. . . . It is not only essential for the protection of the timber but from the recreational standpoint, for there is little appeal in a fire swept area."

To be fair to Carhart, it was not all gulags. The archipelago also known as the Sangres stimulated his concern for stocking rainbows, for protecting the last flock of wild turkeys in the state, and for protecting the doomed passenger pigeon, sighted in the Sangres along with the luckier turkeys. Further, aware of Hamel's overstocking, he saw the writing on the canyon walls for at least some wild/domestic conflicts: "Mountain sheep are found in the high ranges of the Sangre de Cristo Range. At the present time there are many applications coming in to secure this range for domestic sheep. There are certain ridges [on Mount Blanca] that are now the homes of these bands of wild sheep that will best serve the nation by giving sanctuary for the wild sheep and where there should be as full utilization as possible for growing wool and mutton there should be also a plan to protect these wild members of the family."

Inexhaustible on the uses of the Sangres, Carhart continued, "There is a chance in this sort of a camp to teach better Americanization and higher ideals. The great need of the nation is the Americanization of the people of foreign blood now living in our midst."

After a section characteristically called "Points of Attack," he delivered his penultimate stroke, "a long trunk trail on the high ridge of the Sangre de Cristo range that would carry large parties."

Then, in a grand crescendo, he imagined the Sangres as home to a "Great Rocky Mountain Summer University," a consolidation of all the colleges of the state, or "consolidation of the courses of natural science of all of the schools west of the Mississippi and south of the northern line of Colorado. Or there might be a disposition on the part of some one college in this part of the country to establish such a school by itself."

We may smile at such wild eyes, such dreaming. Still, under a similar dreamer, Colorado College has in fact established such a school on the Baca Grant near the old mining town of Crestone, right at the foot of Kit Carson Peak.

So were these simply dreams? Or nightmares? Carhart never lost his dedication to beauty and his confused admiration for the beast. He simply learned his lesson about changing bureaucracies from the inside. He also took his love for mountain glory into private practice. His career highs include the Denver Mountain Parks system, the Denver Capitol Complex, the campus of Denver University, and Colorado Springs' Myron Stratton Home. Far from losing his interest in the Forest Service, he continued to lobby Congress to force the Forest Service to develop the kinds of regional recreational plans he had first conceived for the Sangre de Cristo Mountains.[13]

While Carhart went on to greater things, one of his and Hamel's most enduring legacies is an exemplification of what Tocqueville found so remarkable about Americans: our propensity to form voluntary associations to address local problems.[14] Today's Sangre de Cristo Mountain Council is only the latest in a long line of citizens' organizations dedicated to the Sangres.

Perhaps a closer look at SIPRA will yield some clues as to how citizens of the Sangres might proceed today, especially when we confront novel, unsettling proposals such as my call for a Sangre de Cristo Conservation Trust.[15] With considerable behind-the-scenes help from the Forest Service, the leading citizens of Pueblo formed SIPRA in 1919. Led by the Pueblo Chamber of Commerce, SIPRA "sold" stock in its operations, which were dedicated to developing the recreation potential of the Sangres at a time when Congress was not funding such needs. Astonishing as it may seem, SIPRA raised and spent more than $200,000 on various recreational improvements in the first few years of its existence. Its biggest corporate donor was CF&I, though many other prominent businesses and individuals pitched in.

SIPRA eventually ran into difficulties, both internal and external, with the Forest Service, with Carhart, and with competing interests such as stockmen. But it did not labor entirely in vain. The San Isabel was a hard sell early in this century, after fire, excessive timber cutting, and grossly inflated grazing had compromised the forest's "protection" role and threatened to neutralize Hamel's reclamation efforts. Larkspur and locoweed, the bane of domestic livestock, proliferated. (Even today, the name "Loco Flats" sticks to huge meadows in the Upper Huerfano like a pesky burr, while larkspur blues entire Julys farther north in the Wet Mountain Valley.)

After SIPRA's difficulties, things looked typically glum for recreation development in the Sangres. But the advent of the Civilian Conservation Corps (CCC) saved the day in an ironic way typical throughout the West. When private efforts failed, massive government intervention did the job, whether in predator control, dam building, erosion control, or recreational development, often yielding as much harm as good, especially for biodiversity.

The first CCC camp in Colorado burst into bloom in the nearby Wet Mountains, just east of the Sangres, in April 1933. In August 1934 another camp flowered nearer the Sangres themselves at Gardner, along the Huerfano River. In 1935, swarms of CCC workers terraced hundreds of acres of locoweed-infested, larkspur-plagued acres in the cow-burned, sheep-blasted, horse-harrowed Upper Huerfano watershed. They fenced out Ulus Herard's marauding horses and cattle. Then they planted these areas in exotic bluegrasses and vast numbers of ponderosa pine stock derived from nonlocal sources.

The Sangres-long Rainbow Trail was a result of one of many joint efforts of SIPRA and the CCC. More than one hundred miles long, and lower than the crestline route Carhart envisioned, the trail has served as the chief fire-fighting access for the east side Sangres for half a century now. It also has been a persistent nightmare for anyone concerned about recreational impacts on the Sangres, an ironic match for the adverse impacts on the Sangres' biodiversity of livestock, larkspur, locoweed, and bluegrass.

Chronic underfunding in the face of overwhelming demand—that is the history of recreation "management" of the Sangres. Official files overflow with documentation of this history. Whistle-blowers retire or move on. As long as environmentalists oppose the imposition of recreation fees that work where they are levied, there seems every reason to expect wilderness status to worsen this depressing pattern. In a 1970 history of the San Isabel former Supervisor Cermak wrote, "In effect, the CCC had taken over the duties of the Association." Perhaps with tongue firmly set in cheek, he added, "The basic structure of the first recreation plan is essentially the same used today. . . . The San Isabel never became an important recreation center despite the efforts of the Association, the Forest and Carhart."[16]

Conclusion: Only Connect

Biodiversity in today's Sangres somehow survives the pressure of too many visitors and the neglect of four ranger districts, two national forests, two Bureau of Land Management Resource areas, four Division of Wildlife areas, a National Park Service monument, State of Colorado school section lands, numerous contiguous private lands (some very large), and relatively few National Forest inholdings. Now the advent of wilderness status threatens to tip the balance.

How are the agencies preparing for the advancing hordes? Thanks to the initiative of San Carlos district ranger Cindy Rivera, progress has begun toward interagency cooperation in managing the Sangres. Only recently have all these competing entities agreed on the recreation-related basics, such as standards and guidelines for roads, trails, campgrounds, outfitters and guides, and signing. No one, however, is paying any official attention to biodiversity.

Economists and ecologists converge at this point. The transaction costs of administering the Sangres for biodiversity are insuperably high. The free-rider problems of leaving precious biodiversity resources open without charge to anyone seem insurmountable. Since the national forests in the Sangres have lost money for ninety-one years of their ninety-one-year existence, the case can be made for change. Their administration should be turned over to a locally based nonprofit institution, the Sangre de Cristo Conservation Trust. Using geographic information systems, we can model the effects of multiple management changes in the Sangres, never damaging a hair of what exists today. Using our imaginations, we can dream of a time when we can give biodiversity in the Sangres a chance.

After all, my title reads: "Beauty *and* the Beasts." Allowing for their ecological rarity, we can also determine the value of predators' absolute rarity in the Sangres. Rather than pity predators from a safe distance, we can take Blake's warning to heart. It is we who make biodiversity in the Sangres rich—or poor.

Notes

1. Philosopher/animal trainer Vicki Hearne best articulates the possibilities for mutual respect between humans and animals.

2. Robert Allen and Robert Peet, "Gradient Analysis of Forests of the Sangre de Cristo Range," *Canadian Journal of Botany* 68: 193-201; Robert Allen, Robert Peet, and William Baker, "Gradient Analysis of Latitudinal Variation in Southern Rocky Mountain Forests," *Journal of Biogeography* 18: 123-39; Robert Peet, "Forests of the Rocky Mountains," in *North American Terrestrial Vegetation*, ed. M. G. Barbour and W. D. Billings (Cambridge: Cambridge University Press, 1988), 63-101.

3. John Terborgh, "The Big Things That Run the World—a Sequel to E. O. Wilson," *Conservation Biology* 1 (1989).

4. Aldo Leopold, *Game Management* (New York: Scribners, 1933) and "Deer Irruptions," *Wisconsin Conservation Bulletin* (1933); John Mitchell and Duane Freeman, *Wildlife-Livestock-Fire Interactions on the North Kaibab: A Historical Review*, General Technical Report RM-222 (Fort Collins, Colo.: Rocky Mountain Forest & Range Experiment Station, USDA Forest Service, 1993).

5. Mary Britt, "Building the Elk and Antelope Herds: Dan Riggs' Four Decades of Wildlife Management," *Sangre de Cristo Magazine* (1987).

6. Dave Hoart, Colorado Division of Wildlife. Interview with author. August 1992. Hoart is Dan Riggs's successor. There were 300,000 elk in Colorado in 1993, even after a 1992 harvest of 48,776 by more than 300,000 hunters. Hunters also took 73,995 deer from a population of around 500,000. Licenses cost $20.25 ($150.25 for out-of-state hunters) for deer and $30.25 ($250.25) for elk in 1993. Additional tens of thousands of "antlerless" licenses have been issued, but without making much of a dent in population curves.

7. For comparison's sake, space requirements can be generated for other predators as well. While the managers of fragmented public lands have shown little or no interest, independent biologists have calculated habitat needs for the nearby San Juan Mountains, across the broad San Luis Valley to the west of the Sangres, but joined to them in the north by Poncha Pass. According to their numbers, each wolf would require about 90 square miles of habitat, while a wolverine needs 40, a lynx, 30; a male black bear, 35; a female black bear, 10; and a goshawk, 8. See Tony Povilitis, "Applying the Biosphere Reserve Concept to a Greater Ecosystem," *Natural Areas Journal* 13 (1993).

8. Donald Worster, *Nature's Economy* (Cambridge: Cambridge University Press, 1988), first published by Sierra Club Books in 1978. As Worster shows, Coloradoans owe a debt to the courageous Rosalie Edge for exposing the Audubon Society's bloody hands.

9. Jack Williams, *A Biography of Ulysses V. Herard*, Great Sand Dunes National Monument library, no date. See also Steven Trimble, *Great Sand Dunes: The Shape of the Wind* (Southwest Parks & Monuments Association, 1978); Walter E. Perkins, *Historical Sketch*, San Isabel Forest Preserve, San Carlos District, San Isabel National Forest, 1922.

10. Wildlife Journal, 1922-45, San Carlos District, Canon City, Colorado.

11. Robert Cermak, *Plans Must Be Big and Broad: The Beginning of Recreation Planning on the National Forests*, San Isabel National Forest, ca. 1970. Cermak was supervisor at the time he wrote this piece.

12. Arthur Carhart, *Colorado* (New York: Coward-McCann, 1932).

13. Arthur Carhart, "Passes Over the Blood of Christ," in *Westerner's Denver Posse Brand Book* (Denver, 1946), 183-200.

14. Alexis de Tocqueville, "Of the Uses of Which Americans Make of Public Association in Civil Life," in *Democracy in America*, ed. Richard Heffner (New York: New American Library, 1956), 198-202.

15. See Tom Wolf, *Colorado's Sangre de Cristo Mountains* (Boulder: University of Colorado Press, 1995).

16. Robert Cermak, "History of the San Isabel National Forest," *Pueblo Star-Journal and Sunday Chieftain*, 3 May 1970.

12 . Healing by the Wilderness Experience

Robert Greenway

The wind is frigid as we climb the ridge to greet the sun. "Why are we doing this?" we wonder, as we crawl out of warm sleeping bags into the icy predawn dark. Now, warmed and awakened by the climb, the pale apricot glow brightening to the east, the blue-black of night turning violet to the west, we know we are walking into the main drama of all life, of our particular corner of the galaxy: the return of the sun, the source of all energy. We open up the genetic memories of the thousands of years during which the sun was worshipped lest it might not rise. We have begun our day in balance and delight, and what good it does us we do not know . . .

For all the overcrowding of wilderness areas in recent years, a full "experience of wilderness" has become increasingly rare. Certainly few of us live at the boundaries or near trailheads. None of us lives within wilderness. To get to these enclaves of pristine natural processes we must drive in shiny vehicles for hours on freeways, through endless suburbs and intensively managed farmlands.

Similarly, wanting to discover "truth" about the wilderness experience, we can go there together, or we can cruise around in the cities, suburbs, and gardens of ideas, embedded in cultural forms. And that expresses a first paradox: using culture to understand the experience of where culture is not. Whether we love its qualities or hate its failings, whether we see culture as a monolithic prison or a creative stimulus, we are indeed the mammals with the symbols, the only creatures on the planet (except, perhaps for our domesticated pets) that have become embedded in the processes and products of our mentality.

I am interested in finding ways for those who have felt the wilderness experience, and especially for those whose work is to guide others into that experience, to share what we know without lapsing into incoherent hyperbole or psychobabble, and to lead others on the trail along with us.

I accept that, with regard to the planet's environmental health, we are approaching, or passing thresholds of irreversible damage to many species' (including the human species') chances for "normal" evolutionary unfolding. Whether one picks the local watershed or underground aquifer, regional issues such as massive destruction of forests across many watersheds, or planetary concerns such as overpopulation, desertification, global warming and the like—whatever one's perspective, the crisis is real, severe, and growing.

This of course affects every so-called wilderness left on the planet. From high on the wilderness mountain it appears that our Western civilization is a tide washing up every canyon, pressing into every natural system. Even the rain carries into the wilderness poisons from the machines and factories that drive our civilization. Even while the mental boundary between wilderness and culture grows more polarized every day, damages from cultural processes grow more pervasive and the tide overwhelms the wilderness boundary on all fronts. Nothing slows it—not more designation of wilderness lands, not local and regional growth management laws, not the resolve and litigation of environmental organizations. As the culture wins, the planet—or at least the human species—loses.

This is a second irony we must face: as wilderness diminishes, the desire to have some kind of wilderness experience grows. As many as 50,000 users may hike in on a busy weekend near large western cities (Seattle, Portland, the San Francisco Bay area, Denver, etc.). Camping spots must be reserved long in advance. Latrines appear in backcountry. Solitude is virtually nonexistent, even as organizations of backpackers and other wilderness users vociferously resist any attempts to adopt quota systems or zoning to limit certain types of usage. To advocate the wilderness experience as an essential ground of healing is to advocate an experience of decreasing availability.

Further, where much of the rest of the world's population looks upon wilderness as food for survival, the North American possibilities for experiencing wilderness stand in stark contrast. We must, I believe, build into our understanding of the wilderness experience the idea that though our culture rests on genius and immense creativity and has achieved heights of aesthetic activity beyond description, without a doubt it also rests on the diminishment and destruction of a large percentage of the natural world. In other words, the collective wealth that allows us to enjoy wilderness and many related activities is rooted in our history of having consciously and purposefully exploited wilderness to the point of its destruction over much of the planet.

It is also important for points I want to make below to state my belief that the causes of the environmental crisis are not as well understood as many pundits would have us believe. Leading candidates are overpopulation, various political systems, various economic systems, abstractions such as greed, and mass activities such as consumption. There are only a few, astonishingly, who seem to notice that behind the abstract humans who are behind the statistics and abstractions are very specific mental processes—the underlying matrix of culture—and that given the recency of cultures in the life of the planet, something might be going very wrong with the most fundamental way in which we create reality and process information.

Whatever the causes that emerge as symptoms in the news, I believe that such causes are carried by our culture, in language, stories, myths, and layers of assumptions, and that this culture is more pervasive, co-optative, and thus perhaps more damaging and difficult to change or heal, than even the most alarmist experts believe. Thus, I conclude, painfully—because I've said the opposite hundreds of times—that leaving the culture is a romantic and unrealistic notion. And yes, it's a notion that's been around since antiquity—from the beginnings of urban culture there are accounts of people escaping to the country for peace and quiet, perspective, resolution of difficult choices, empowerment, vision, realignment with deities, and the like. There are accounts of such people returning more able to function more powerfully, or more viciously, or with different assumptions about what culture should be. And of course some didn't come back at all. Something happens to people who leave the overt reinforcements of the culture, but it is too simplistic to say that the culture falls away and "nature comes in and heals."

Of course we leave culture physically, but if we're culture bound—locked into a voracious web of continuing reinforcements that penetrates into and is supported by mental processes—then what is it that changes psychologically through the wilderness experience? I believe our inability to answer this question reflects our meager understanding of both the psychological processes underlying cultures and the interactions between mind and nature.

Of the dozens of distinguishable mental functions (organized into what might be called "modes of knowing") of which humans are capable, different cultures tend to emphasize one or more over others.

Whereas Native American cultures might reinforce the ability to tune into the simultaneously multiple functions of systems, a perspective necessary for tribal life and for intimacy with the natural world, Western cultures might emphasize ranking of perceptions into hierarchies, or objectifying and isolating distinctions, or stopping time in order to examine things. Native cultures might emphasize the kind of all-encompassing awareness required for hunting; Western cultures might emphasize focus, the narrowing down of perception so useful in certain kinds of problem solving.

Of course this is greatly oversimplified, for we know that (especially in recent history) styles of cultures merge, cultures build up certain self-feeding mechanisms over time, modes of knowing change throughout stages of life, and gender differences (whether intrinsic or products of socialization) profoundly affect how we see the world. In huge and complex cultures like our own that contain vast areas of shared assumptions, subcultures form around certain shared cognitive styles, fueling belief systems and ideologies. The various media by which information is conveyed shape what is seen, known, and believed.

The most notable cognitive emphasis of our culture is what philosophers call *dualism*. I think it is an important and revealing doorway into the psychological changes we see in those staying long enough in the wilderness to leave as much of their cultural props behind as possible.

By dualism I mean its radical form, a complete divergence of realities: a person is aware of differences, but sees the differences as existing in different realities ("man's ways are not God's ways," "the stuff of mind is different than the stuff of bodies," "the natural and the supernatural," etc.). It's as if we've taken the marvelous (and evolutionarily crucial) mental function of distinction making and pushed it until distinctions become disjunctions. Thus, normal perceptions of figure-ground, or contrasting linked differences (dark and light, mind and body, good and evil) known as dualities become dualisms when they split into different realities (i.e., the devil's realm of evil, God's realm of the good). Although people full in the throes of dualistic mentation are usually unaware of it, and the dualistic mode is supported by very deep belief systems and constant cultural reinforcement, there is the alternative belief, perhaps on the rise in this supposed dawning of the "ecological age" (and perhaps supported by a never-completely-extinguished intuition of wholeness, or the under-the-culture continuing presence of what are intimate "women's mysteries") that dualistic splits in reality are illusory—a massive cultural con game.

This is not necessarily as comforting an awareness as it might at first seem, for along with the perception that dualism just happens to be our particular mode of knowing is the perception that this mode is completely intertwined with our deepest levels of cognition and with our institutions, mythologies, and value systems. As John Dewey pointed out more than sixty years ago, what is American individualism but dualism writ large?

Dewey also so perceptively pointed out (well before Gregory Bateson picked up on the same theme) that once a genuine dualism is established the dualistic mode spreads like a disease: mind and body, matter and spirit, visible and invisible, all the chauvinisms so plaguing us—and, of course, the increasingly tragic dualism we give our attention to here, the split between human mental life and nature, between culture and wilderness.

As with all dualisms, it matters little to just define our situation as dualistic if our fundamental mental processes—those beyond our ability to change—remain dualistic. Thus, as I said earlier, the diagnoses of the crisis of our relationship with nature ("the cause is greed," "the cause is capitalism," "the cause is the Judeo-Christian heritage," etc.) provide little or no relief so long as our personal, fundamental mental processes remain bound in the dualistic mode.

Ideas divorced from experience tend to be split off from the whole. This means that because our physiological processes remain embedded in natural systems and our individual and collective mental life thinks and believes and acts as if it is somehow independent from natural systems, and if our mode of knowing is capable of spinning out ever more amazing inventions, then it becomes relatively easy to assign natural systems to an inferior status, something to objectify, study, master, conquer, exploit—or destroy.

As stated, when one is in a dualistic mode, one side tends to become perceived as the whole of reality, while the other is either out of consciousness or something of which one is only vaguely aware.

Even if we are aware that our thinking is dualistic and so we are cut off from the awareness of chunks of reality—aware that our separations are illusions and that our physical processes are not separated—so long as our fundamental mental processes remain dualistic we will not have the benefit of consciousness of the flow of the world around us. Our knowing will perhaps be acute, laced with accuracies gleaned laboriously from the efforts of science, but the full picture—the emotional, experiential, fully interactive sense of connectedness—will elude us. And, I believe, haunt us.

To some degree, all that we know will be distorted. The emotional intimacy possible only through fully accurate knowledge will be unavailable to us. This, I believe, is our situation with regard to nature, the reason that our diagnoses with regard to the human-nature crisis seem so limited, and the reason that, despite the severe social contradictions and paradoxes, the experience of wilderness holds out very special hopes and insights. We can see the incredible double bind: it is said that lack of intimacy with nature makes us crazy; and it is said that our culture is pathogenic with regard to nature. Thus, to find sanity, it would seem sensible to leave culture, and immerse oneself in nature.

It would indeed seem obvious that getting out of the pathogenic culture would be a prime means of reestablishing intimacy with natural systems. Wilderness seems the obvious locale for healing, because it is where culture is not.

But as countercultures and back-to-nature movements have vividly shown, both physical and mental removal from the seductions, benefits, and necessities of our massive, pervasive culture are very hard to come by. One tends to live in parasitic relationship with Western culture even as one basks in the backcountry meadows of the wild western homestead surrounded on three sides by national forests. From the gas to run the generator, to the propane to run the refrigerator or the high-tech batteries to run the stereo, the appetites rooted in our culture's mythologies of human superiority and autonomy are virtually impossible to assuage.

Similar dynamics surround the growing movement to utilize the wilderness as context for healing. Most of the literally thousands of organizations offering wilderness excursions approach the wilderness as a cultural idea. What we tend to love—and I say this after having led wilderness groups for more than twenty-five years—is what we project onto the experience: a kind of sexy vitality, where bears are sweet and are likely to say dumb, funny things in a deep voice; where the noble stag—"That's your father, dear"— stands high on the knoll facing down civilization off in the distance.

What we see in much of our advocacy of the wilderness experience is our own peculiar culture, written on the presumed empty blackboard of wilderness, the "wasteland" used by the air force for bombing and low-altitude combat practice. One wonders, sitting around a tiny middle-of-the-night fire deep in the wilderness, whether the dumping of the pain and poisons from our lives in a culture broken from nature is really that much different than the military use of the same territory; whether using the wilderness for therapy is really different than the other forms of exploitation of wilderness throughout our history.

Leaving that question aside, all I know is that if the wilderness experience is to transform, one must enter it openly and fully psychologically; one must be able to loosen the dualistic processing of the culture and find communion with—as opposed to projection onto—the realm of natural processes. One will experience incredible aesthetic pleasure, and must not drag it out of the immediate experience into the cultural categories of "pretty" or "scenery" or into such abstractions as "health" or "life everywhere"—for what one is experiencing is not just nice young sleek plants and critters everywhere, but death everywhere as well.

True, to immerse oneself in the wilderness with a sense of opposition to "the culture out there" is another form of dualism. Over the years I have come to accept this as a necessary strategy. I do everything I can—silences, no mention of the "other world," no discussion of movies or media or one's fantastic equipment—to leave the other world behind. Like summer camp residents, we even assume new names. The food is tightly controlled, "just enough," very nutritious, but light, with little variety. No books, paper, pencils. A few tools, the minimum equipment required for safety.

My theory behind this is that, to come out of a full-blown dualism, one must experience both sides, both realities. Experience! Not think about it, or adopt some ideas and stances, but really experience. So, assuming our culture to be in opposition to nature, now is the time when we have the luxury to live within nature and are more or less in opposition—not to culture in general, or even to the sum total of Western culture—but to those aspects of our culture that are killing the very natural systems in which we are immersed.

Of course we writhe over the inconsistencies and contradictions—the long drive to the trailhead, the high-tech equipment, and the like. And there is often deep suffering over the more overt news of the various impending environmental disasters, and personal wounds and suffering (born of environmental disjunctions or social wounds as well) projected onto the wilderness ("The trees are weeping for the planet!"). All of this may, at the outset at least, be useful, *healing*.

But over the years I have become interested less in emotional displays of empathy with trees and animals that weep than in the underlying psychodynamics behind the evidence of destruction and imbalance for which nature is purportedly expressing human emotions.

I am in mourning, very frankly, for the opportunities for experience that will not be available to my grandchildren, and for the incredible capacities of human life, which we glimpse here and there in the high points of culture though for the most part we view the waste and destruction of addiction (addiction as the false promise of fake solutions, the trading of long-term reality for immediate pleasures).

My goal has become (over a long, long apprenticeship) a fiercely practical one: to have whole days, even weeks, apart from the barrage of reinforcement of the dualistic mode, and to experience whatever modes of knowing, whatever patterns of reinforcement—exist in those few places "where culture is not."

Simple things: I rediscover my digestion by putting my shit in a hole in the ground every day. My senses open up—my safety depends on it. My skin becomes taut from the ice water of the rivers and contains my interior in a different way. My signals are picked up in my consciousness, instincts are aroused, primal connections with water, fire, air, earth emerge and seem familiar. My boundaries seem to become permeable—between waking and sleeping, consciousness and unconsciousness. What previously appeared to be "my" tightly contained mind seems to interact with the objects of my perception. I do not lose my sense of the subjective, or of being subject, but my "self" obviously interacts with the ground out of which it had been separated. Yes, it is home, filled with incredible beauty, death, destruction, constant change. I have entered into a seething, changing, balanced community that was doing just fine without my awareness of it.

My task in the wilderness, and what I try to facilitate, is to learn how my awareness can open without intrusion. It is like becoming worthy to be chosen as a lover, and I want to be chosen. The ideas waft in and out, days slip by, the simple tasks of living become fully occupying. Soon the differences between images and ideas—that demarcation line at which experience becomes acculturated—becomes apparent. I am moving down a river, immersed in my own safety, fully open with every sensory mode . . . experiencing. It becomes a watershed—a cultural idea, but a useful one, and it easily remains connected with what I'm experiencing. The watershed begins to appear elegant, stupendously balanced and perfect—a cultural idea, an abstraction, even—yet it remains embedded in the constant inflowing perceptions, my experience.

As the time to return to "civilization" draws nearer, and I stay with the open awareness, I can come to notice when ideas begin to break off and become isolated from experience—become abstractions. I create channels, like software, among experiences, images, and abstractions so that, however these wilderness ideas might be used in the "other world," the flow from nature into culture will not be broken.

What I term "healing" or "transforming" is when the idea of our separation from (and thus superiority over) nature appears not as an all-encompassing reality, but as an interesting idea, a game I will have to play in that other world.

We are almost packed and ready to leave the wilderness camp when I notice Lucy, one of our high school students, is missing. Her backpack is missing too.

I think I know what is going on. I have seen the bitterly unhappy life she lives "in the other world," and I have seen her flower into a powerful, graceful, and often ecstatic young woman after two weeks in the wilderness.

As the day to leave neared she had been looking increasingly bleak and I had been concerned about her.

I jogged up the canyon and turned up a side stream to a high meadow above our base camp, a place near where we had climbed every morning to watch the sunrise, one of the most pleasing spots I had ever experienced in any mountains.

Yes, there she was in the middle of the meadow, her back to me, facing the sun. I walked up to her. "Come on, Lucy," I said. "We're leaving." I looked at her face and the tears streaming down. I will not forget her words:

"Just tell me one good reason why I should go back."

I thought of what I knew of her life, her abusive parents, the ridicule of just about the entire school, few or no supporters among the teachers. She was a misfit in that world with her thing about wearing shoes, her passion for small creatures, her almost total silence there.

I could only remind her of the promise I had extracted from everyone at the outset, to "touch the cars" at the end of the trip before being freed of the rules and commitments of the class.

Later I watched as she and the other students integrated themselves back into the high school—starting as shining beacons standing out in the crowded halls of the high school between periods, then dimming and paling until, within two weeks of their return, they were indistinguishable from their fellow students.

In the earliest phases of my guide work the return from a wilderness excursion was of little concern. People by and large were glad to have a hot shower and get back to "normal." But later, as I began using forms and processes to speed the loosening of the grip of the culture and enhance and deepen the opening into wilderness, upon return many of the participants would suffer severe depression or other symptoms of disjunction.

I began realizing I was inadvertently creating a drama of profound conflict, a cultural conflict with nature brought to the surface through the wilderness experience that could seriously disrupt, if not devastate, one's life.

Many issues arose: just how extensive a transformation of one's reality processing can take place in a two- or three-week immersion in wilderness?

Can a deep-level psychological immersion in wilderness be integrated into a culture that is fundamentally antithetical to wilderness? Must these qualities and benefits of the wilderness experience be destined inevitably for depression and conflict?

It has become clear to me that much of what happens in the wilderness and on the return depends on the assumptions made by the excursion's leader. It is my opinion that very few wilderness trips cross the wilderness boundary psychologically. Most people, leaders or participants, play out their cultural ideas of wilderness, so the trip enhances their power as defined by the culture, or their ability to integrate into the culture. Such excursions look on the wilderness itself as kind of a theme park, an experience for which one pays an entrance fee and expects a certain amount of adventure, or weight loss, or improvement in a relationship; plenty of gourmet freeze-dried dinners; some campfire songs like in the movies; and warm snuggling in the $350 down sleeping bag with the Gore-Tex outer shell.

I believe it need not be this way. I believe that we can learn to be fully connected with natural systems and that the wilderness is a good place to learn this, and I believe that there are ways to integrate such experiences anywhere without dominance of anti- or nonnatural cultural forms. But it's a tall order.

I think the depression following wilderness experiences is an important clue. However depression is defined, it is quite clear that it involves a lowering of available energy (thus, the feeling of loss). There is always a sense that one is below the threshold of energy necessary for well-being or to "get work done" in the world.

If the wilderness and the connections with natural systems (including one's own body) greatly raise the energy level, what happens upon return? Obviously those connections are broken or blocked in the human-created world to which we return. It is not safe to keep one's sensory processes open in the city.

The assumption I am making, based on return interviews of more than 700 participants in wilderness excursions, is that the myriads of connections available in urban life, loud, flashing, colorful, symbolized as they are, do not begin to provide the same level (let alone quality) of energy that is available in the wilderness. Further, it takes energy to keep sensory processes closed, limited, or as selective as they need to be in urban life. So there are many losses following a full psychological experience of connectedness with nature and no amount of Prozac, caffeine, chatting through the wires of the

Internet, or racing around behind the equivalent of several hundred horses seems to assuage it. In short, it takes a huge amount of energy to maintain something as unnatural as a dualistic way of processing information—even when that information is half-digested, canned and packaged and the processes to deal with it are provided for us.

A solution of sorts appeared to me, arising out of a mixture of studies of systems, consciousness, neurophysiology, and Buddhist writings and practice, combined with a return to the level of musical involvement of my childhood. It was a solution so simple that I had been stumbling over it for some time.

I came to realize that one didn't need to replace connections made in the wilderness with aspects of our culture that are destructive of nature. In other words, to see nature and culture as an either/or situation is simply to maintain the kind of dualism I was trying to heal. What I needed was to learn how to maintain connections with nature anywhere. I began realizing that it was crossing a wilderness boundary to maintain awareness of my heartbeat, my breathing—to allow all my bodily functions. I began realizing the underlying "given-from-nature" aspects of human (and highly acculturated) activities such as eating, gardening, sex, child-rearing, and dying. I began to see that on every front, in every crack and cranny of our culture, the drama between nature and culture is being played out—abortion, genetic engineering, letting people die natural deaths, and so forth—and that the transition between wilderness and culture occurs virtually all the time!

What isn't occurring in the city, of course, is the awareness of the prevalence of nature, its underlying presence in every aspect of our lives, no matter how far we think we've been able to distance ourselves.

I also found in music a bridge reaching from culture back to nature, just as I found in growing my own food a bridge from the soil to our table. And most helpful of all, a serious practice of meditation became the same as the wilderness experience. I began requiring meditation and yoga of my students as part of our preparation for a wilderness trip, during the trip as part of our daily practice, and then after return to the city. Depressions upon return almost completely ceased. And from this I can only conclude that the experience of wilderness and the serious practice of meditation are identically about the reclamation of awareness, and such awareness is what we call "healing," since it seems to have the capacity to make permeable the walls created by a dualistically dominated mode of knowing.

The final irony here, then, is that what we learn from the wilderness experience that can heal our relationship with nature is not exclusive to wilderness—although the wilderness has been a wonderful ground in which to learn it, and I'm profoundly grateful for having had the privilege of being able to spend so much time there. There are human activities that no matter how acculturated cannot hide their roots in instinct, soil, hormone, or sensory event.

This is the time in history when we cross the line from an isolated epistemology to an epistemology that reconnects our full consciousness with our full immersion in nature, and when we recreate communities, institutions, and new forms of culture that contain the fullness of both.

This is the time in which we can come to realize, moment by moment, that each breath taken into the depths of our lungs exchanges certain molecules with other molecules, with differing electrical charges that can be measured. And this is the time when we can come to realize that each breath is a communion of the immense neurophysiology of our minds with the wilderness mind of our blood; the time when we can learn to know, fully know, without distortion.

Photograph by Adam David Clayman. By permission.

13 . Urban Wilderness

Andrew Light

Wilderness began as *the* term that marked out the cognitive space that was not civilization. The wilderness was the jungle, the wild place, not fit for human habitation except for those who were not really fully persons—the barbarians. The wilderness marked not just the geographic boundaries between human settlements and wild nature, but also a cognitive boundary between the civilized explorers and the "savages" they encountered. The wilderness was, in short, the mental and physical boundary between humans and the radical/racial others.

The ideological transformation of wilderness continued by colonizing new territories as wilderness areas. This movement can first be seen at the beginning of the twentieth century in America, where metaphors of the growing "wilderness in the cities" were used as a rallying cry for social reform against exploitive labor practices. In this context reformers suggested that if the wilderness was where we civilized people were not supposed to be, then inhuman physical conditions in our civilizations were intolerable and had to be changed.

Today, however, we have come full circle back to the use of wilderness and metaphors of the wild to describe those who we civilized humans would mark off as separate from ourselves. This new urban wilderness is the legacy of an older conception of wilderness that is now in retreat: the description of wild nature.

I

At least two clear conceptions of wilderness are identifiable throughout Western history, which may be analytically separated as the *classical* and the *romantic.*[1] The classical view places wilderness as something to be feared, an area of waste and desolation inhabited by wild animals, savages, and perhaps even supernatural evil. In this conception, human society is the standard by which the world is measured, so conquest over the nonhuman areas, the wild areas, signals a form of human achievement, "a victory over the dark forces and a measure of social progress."[2]

The romantic view sees wilderness as an untouched space that human contact corrupts and degrades. In this conception the wilderness is a place to be revered, a place of deep spiritual significance, and a symbol of earthly paradise. Humans go into the wilderness for spiritual cleansing in the uncorrupted landscape. Meditation on the wilderness is encouraged as a path to mental clarity. The history of human encroachment into the wild represents a steady fall from grace rather than a victory over dark places.[3]

The roots of the romantic view are older; however, the classical conception of wilderness was dominant in the West up to the last two hundred years, when the romantic idea began to gain ground.[4] Today, of course, we tend to think of the wilderness as a fairly innocuous place in terms of its danger to man. In the United States we have institutionalized (e.g., through the Wilderness Act) a sense of the wild as something that needs to be set aside and respected so that it can serve as a conduit for beneficial nature experiences. Behind such attempts is at least part of the romantic ideal: wilderness is the form of nature that has remained close to its pristine state, meaning that it has not been corrupted by human intervention.[5] Many theorists have attempted to form an ecological ethic around the romantic idea of wilderness and have claimed that experience with the wild results in a special understanding of the relationship between humans and nature.[6]

What I am interested in here, however, is how the classical conception of wilderness as a place of great danger has persisted over time. The idea of an unknown evil at the edge of civilization still haunts us. To understand why, we need to know more about the classical idea.

The classical view to a large degree revolves around what I call a *cognitive dimension* of wilderness. Here I mean the sense in which the wilderness is a place that is always marked as the realm of the savage who is thought to be cognitively distinct from the civilized human. The savage is always marked as the thing that we outside of the classical wilderness, we civilized people, are not. We are superior to the savage simply in not being the savage and part of what makes us not savage is our possession of reason, our control over our passions. The savage is supposedly subject to his passions and is in fact driven by them to the point where he may not escape the wilderness. But the escape from the wilderness is not merely a physical escape; savages who leave wild nature are still wild because of the "cognitive" wilderness within them. The cognitive dimension of wilderness thus refers to the wildness *within* the beings who are part of wild nature; it is not the mere physical surroundings but the claim of those surroundings on the mentalities of its inhabitants. To be able to mark off the wild from civilization therefore does not require any specialized knowledge of wild places with respect to their mere physical descriptions, but only a strong sense of the difference between us and them—between we civilized rational agents and those uncivilized, passion-filled savages.

The classical view, with this important presupposition, was the dominant view of the wilderness that traveled from the Old World to the New World and shaped the early Euro-American perceptions of Native Americans. John Smith described New England in the seventeenth century as "a hideous and desolate wilderness, full of wild beasts and wild men."[7] So uncontested was this assumption of a cognitive distinction between civilized and uncivilized in early American history that different thinkers developed radically disparate views of the potential of civilizing various kinds of savages without ever questioning the grounds for their debate. While some believed that the African "savage" could be civilized (with appropriate guidance) outside of their wild jungles, they held for various reasons that other savages—most importantly Native Americans—were beyond hope. Others, like Jefferson, argued the reverse.[8]

One way of looking at the extremity of this situation is to recognize that the classical wilderness becomes a stand-in for a place in which the civilized do not want to go—the result of spiritual despair, a replacement for Hell. Certainly one cannot help but think that the cognitive wilderness was, at least for the early Puritans, a projection of their deepest fears for what they could become in this new land if they were not careful. As Short says, "The wilderness becomes an environmental metaphor for the dark side of the human psyche."[9]

The classical position can be summarized as emerging out of three related (and often overlapping) theses, that is, three characteristics that must be present in some form for there to be a wilderness of the classical kind:

(P1) *Separation.* Since wilderness is bad, evil, and cruel, it must be separated from humans—it must be marked off as distinct and kept out of civilized spaces.

(P2) *Savagery.* The inhabitants of wilderness are nonhuman beasts and are accordingly demonized and vilified.

(P3) *Superiority.* In contrast, civilized humans and civilization may be celebrated in their successful separateness and triumph over wilderness as virtuous and superior.

I will refer to these theses throughout this discussion in order to test the existence of classical wilderness at different moments in history. The idea of the classical wilderness is, of course, constantly going through revisions, so the relative existence of these characteristics, as applied to certain spaces, may indicate a kind of wilderness in transition. For example, as Euro-Americans began to eliminate the threat of Native Americans in the 1800s the relevance of P2 diminished for a time, but for some the wilderness still existed as a place inferior to white civilization (P3).

Certainly classical wilderness is a cultural construction with an identifiable history,[10] and as an antidote to this view the romantic conception is likewise a cultural construct created for a certain purpose. Surely the wilderness (or, as I prefer to think about it given this genealogical account, "wild nature,") does contain a dimension that is scary and harsh, the experience of which helped to form the classical ideal. And wilderness can still be a place of personal transcendence, as the romantics claim. But the reality of the wild must lie somewhere between these two poles of terror and bliss: yes, wild nature is savage *and* it is also beneficent, but certainly not exclusively either all the time.

One familiar way of beginning such an argument is to look at world views that do not contain either of these cultural constructions—that is, to examine coherent views of the wild that do not seem to rely on an extreme vilification or celebration of wilderness.[11] What is wild nature outside of either of these Western paradigms?

It is well known that the original inhabitants of North America did not have a special idea of wilderness as something set apart either hostilely or romantically.[12] Interestingly enough, no word in any indigenous North American language has been identified as meaning what Europeans or Euro-Americans take to mean "wilderness."[13] This may strike some as odd, particularly since Native Americans were supposedly the "savage" inhabitants of the wilderness when Europeans first came to North America. But I think we should not find this so surprising. In either the classical or the romantic view, *wilderness* does the work of a "natural kind" term: It picks out something outside of us and inheres it with a certain definite range of qualities that are uncontested inside of the language system that accepts the term. In either the classical or the romantic description the term refers to, among other meanings, something outside of ourselves, something identifiable, separate, foreign, and distinct.

But for Native Americans, no such term was needed. The term *wilderness* does not pick out anything separate or distinct or alien in any way for them. It is not a natural kind term at all because it does not pick out anything particularly distinct in the world physically or even cognitively. Words for hills, streams, plains, or forests are enough to identify the wilderness around. If these areas are home, why should they come up with a term to signify this area as imbued with alien properties? A classical conception of wilderness, in contrast, is a good idea of wild nature to have if you want to separate new colonial settlements in a new place from the world around you (and, of course, justify a suppression of the "savages" in that hostile world). In that case something is being identified as separate and distinct and deserves in some sense a term to designate it as containing certain properties.

Even with this account I think that more is needed by way of an explanation of why Native Americans do not refer to wilderness as Europeans did upon their arrival to the new world. Certainly it is possible that indigenous North Americans could have a word for something approximating a special sense of "wild nature" as distinct from settled areas, and we may have simply lost this term over the course of the steady obliteration of Native American culture. And it is possible that the idea of the absence of the wild for indigenous peoples is itself a cultural construction romanticizing their disconnection from Euro-American culture.

Recent work by ethnobotanists suggests that the roots of the "absence of the wild" for Native Americans may have a more coherent material base. The argument is essentially that the physical stuff we in the United States refer to as "wilderness" (both when Euro-Americans first arrived and today in statutes such as the Wilderness Act) was for indigenous peoples cultivated land. For example, in California, what early explorers found to be a rich, impressive, and daunting "wild" landscape thought to be a "natural untrammeled wilderness, . . . was to a large extent actually a product of (and more importantly dependent on) deliberate human intervention."[14]

The "wilderness" for native Californians did not exist because it was not separable (P1). It was a basis for the material culture that was the foundation for everyday life. Methods used to change the landscape were unrecognizable from the point of view of the tradition of Euro-American agriculture (e.g., controlled fire, seed beating [for seed dispersal], transplanting of shrubs and small trees, construction of ditches for irrigation, pruning and cropping, and weeding and tillage of specific plant communities). This process was so extensive, according to Kat Anderson and Thomas Blackburn, that when it stopped the "wilderness" began to recede:

> The domesticatory process here seems to have reached the point where important features of major ecosystems had developed as a result of human intervention, and many habitats (e.g. coastal prairies, black oak savannas, and dry Montana meadows) were deliberately maintained by, and essentially dependent upon, ongoing human activities of various kinds. . . . the various plant commu-

nities that early European visitors to California found so remarkable were largely maintained and regenerated over time as a result of constant purposive human intervention. When that intervention ceased, a process of environmental change began that led to a gradual decline in the number, range, and diversity of many of the native species and habitat types that once flourished here.[15]

Native Americans were not idle hunters and gatherers who left nature in a pristine state open to fear or reverence. Instead, the "wilderness" was a part of everyday life maintained as the urban public sphere was and is for Euro-Americans.

The relevance of this argument for our discussion is that it greatly strengthens the claim that the classical and romantic conceptions of wilderness are cultural constructions and puts it on a more concrete foundation. Particularly in the case of North America, the classical and romantic conceptions of wilderness placed an ideological overlay on the land. Perhaps beyond that, what was being referred to, the physical stuff itself, was not wild nature at all. The "wilderness," however you want to look at it and whatever properties you want to give to it, was in a strong sense never really there. What was being referred to was separable only by comparing means of cultivation and evidently not as a "wild," untouched place. Given this position, the word *wilderness* (in either manifestation predominant in the West) cannot be a natural kind term, nor can it be a designator of some condition in or state of the world. The term is simply a powerful metaphor that primarily describes how certain people project themselves into and as opposed to the world around them. In this sense, the cognitive dimension is perhaps the only coherent dimension of the term. While we have good reasons enough to doubt the existence of the physical referent of the term *wilderness,* the cognitive referent is still importantly alive no matter the status of external "wild" nature.

Though this conclusion may seem startling at first, I think it is what we need to make sense of the evolution of classical wilderness since the early American colonial period. For as the romantic idea of wilderness gradually came to replace the classical idea and Euro-Americans found that nature was no longer an evil place, the classical ideal still persisted in some rather unusual areas. I want to suggest that the classical wilderness in its cognitive dimension was simply transferred to the cities. The physical referent of classical wilderness became the urban space as wild nature became less daunting in the face of stronger and more secure civilization. Even though the evil or untouched natural physical space picked out by the term was never really there—or at least is no longer there (since after all it has been conquered)—

what is "out there" still, projected as the cognitive wilderness, is a sense of fear of the "savage space" and its "savage inhabitants."

II

The notion of wilderness has shifted not only between the poles of the classical and romantic views, but also as a reference to nonnatural physical spaces. One of the first examples of this can be seen in the attempt to take advantage of the metaphoric power of the classical view at the turn of the century by American social reformers who used the imagery of the "urban wilderness" as a motivation for their reforms.[16] Aside from numerous uses of this kind of language in reformist periodicals of the time (particularly the socialist newspaper *The Appeal to Reason*),[17] the most famous attempt is surely Upton Sinclair's 1906 *The Jungle*, which documented the abuses of recent Eastern European immigrants in the meat-packing industry. For this contribution Sinclair may be thought of as in some respects the father of the urban wilderness.

In the imagery of the "jungle" we see an attempt to argue for social reform based on the power of naming a space in which humans are trapped in a new wilderness. Capitalizing on the classical conception of wilderness as an evil, destructive place unfit for human habitation (P1), the argument is that where we can recognize people as uniquely human we have a moral obligation to "save" them from the wilderness. The wilderness is no place for those who are inherently superior because they are civilized (P3) and hence in some way more virtuous than their surroundings.

The text of *The Jungle* is full of the language of the classical wilderness. References to "Packingtown" (the home of mainly Eastern European immigrants working in the Chicago stockyards) as wilderness and as jungle are numerous. Even more interesting are the references to the predominance of this classical wilderness. Sinclair has us watch the protagonist of the story, Jurgis, go through all manner of criminal abuses, and consistent with the bewildering (P1) power of the wilderness as corrupting influence, he can never see the harm around him until it is too late. To Jurgis, the unnatural and natural are blurred by the power of the urban wild. Looking out at what Sinclair clearly takes to be the chaos of an urban wilderness wasteland, the innocent reflects:

> All that a mere man could do, it seemed to Jurgis, was to take a thing like this as he found it, and do as he was told; to be given a place in it and a share in its

wonderful activities was a blessing to be grateful for, as one was grateful for the sunshine and the rain.[18]

Sinclair appropriates the classical language of the wilderness to describe the *urban* wilderness, and in doing so adds a new cognitive dimension to the confusion and corruption that breed injustice. A new face is put on the old evil: savage Indians are replaced with savage capitalists (a reserved application of P2); threatening dark landscapes are not the creation of the supernatural but the result of the naturalization of an exploitative labor relationship hidden behind an industrialized factory environment. In making this move, however, Sinclair does nothing to subvert the classical designation of wilderness as uncivilized space. In fact, the wilderness must remain classical for him in order for it to do the political labor he desires. The purpose of calling this "wilderness" is to get us to acknowledge our complicity in the construction of a dangerous physical and cognitive space. Sinclair wants readers to be repulsed at these conditions. Like the early Euro-Americans who were stimulated by the metaphorical power of classical wilderness to build a new civilization as a base from which to attack the danger at its borders, Sinclair wants us to attack the poverty and savagery of the urban wilderness out of our revulsion.

Other writers also have made the case that cities more and more take up the sense of the classical definition of wilderness. "The big city," says John Rennie Short, "is now the modern equivalent of the medieval forest populated by demons." Henry Miller describes New York as a "mad stone forest."[19] At least with Sinclair and the other socialist urban reformers, the argument for an urban wilderness was made out of a concern for its most abused inhabitants. We are not supposed to think that Jurgis and all other poor inhabitants of Packingtown are vilified, demonized savages (P2) merely because they are in the wild; rather, they are people in the wrong place whom we should try to save from their wage-slavery in the yoke of savage machines.

The transformation of urban space into wilderness through a new reconstruction of old themes thus subverts the original physical dimension of classical wilderness while retaining the overall cognitive meaning of the idea. After all, as long as the idea is still around and recognizable, why not put it to good use? The problem is that today we are coming back to a redefinition of urban wilderness as a classical wilderness with the original cognitive content present in the three original theses that I suggested earlier. Instead of separation (P1) for the purpose of beneficial reform as Sinclair desired, P1-P3 are proposed for one of the more insidious purposes of classical wilderness, to justify vilification. And just as we heirs of the romantic view of wilderness, or at least of an anticlassical view, can see a real malicious and damaging potential in the classical view of natural wilderness, we can see potential harmful effects in the construction of a postsocial reform urban wilderness.

III

Los Angeles (where ironically Upton Sinclair was arrested in 1921 for reading the Declaration of Independence in public) is one of the premiere urban wilderness areas in the United States. We may even think of it as something of a "wilderness preserve," maintaining as clear as possible a separation between suburbs and inner city racial minorities trapped in concrete jungles. Recently, we have seen modern-day, twisted, Kiplingesque stories of forays into this wilderness such as the film *Falling Down*. In this fable an L.A.-area defense worker (played by Michael Douglas) decides he cannot take the place or the people anymore and "valiantly" clears out the jungle by force. Interestingly, the filmmakers claim that the tale was supposed to ultimately be a critique of such rash bouts of obvious racism, and was narratively constructed so that by the end audiences would see the error of such alienating attitudes toward the inner city. However, the film became very popular as an "accurate portrayal" of the urban wilderness and audiences closely identified with this vision of a righteous inner-city crusader. Later, American talk show hosts were faced with the problem of trying to explain why people thought the hyperbolic Hollywood representation of life in the city was accurate and why the film was succeeding for all the wrong reasons.

Against the background of the construction of the urban classical wilderness, it seems fairly clear why people were confused about the public reception of the movie. In Sinclair's account urban space is a wilderness, no doubt in the worst metaphorical sense of the term (which for me is really the only sense), but the unfortunate inhabitants are not savages—they are, in Sinclair's subversion of the tradition, deserving of recognition as humans who need our help. In some sense this argument recognizes the constructed metaphorical quality of the wilderness designation while at the same time capitalizing on its semantic history to motivate people for a good cause. But without this motivation to use the term for such a purpose (for example, once the fervor for social reform has passed), all that is left over is the designation of the inner city as a new classical wilderness. The legacy of urban space as a classical wilderness in this sense (devoid of its reformist intentions) is what persists today and what makes it easy, possibly too easy, to swallow the picture of a hostile urban wilderness in *Falling Down*.

But of course the urban conditions in this specific case, culminating in the early 1990s in the rampant turmoil of the Los Angeles riots following the first Rodney King trial, makes the designation of the city as wilderness easier. My concern is that reading these episodes as "urban anarchy" that grabs national attention and persists in everyday life in L.A., is too often informed by the background of the urban wilderness as the most important inheritor of the classical wilderness metaphor. To see this we need to examine how the three theses of classical wilderness can be found in descriptions of inner-city Los Angeles, and how they may in fact shape the material culture of the city.

P1, separation, is not too hard to document in the case of L.A., and has in some ways operated as one of the dominant presuppositions of urban growth and renewal in the area. We can see this thesis at work in the dire descriptions of urban life (not specifically commenting on Los Angeles, but certainly with this city in mind) in the final report of Richard Nixon's 1969 National Commission on the Causes and Prevention of Violence: "We live in 'fortress cities' brutally divided between 'fortified cells' of affluent society and 'places of terror' where the police battle the criminalized poor."[20]

Mike Davis's *City of Quartz*, a most impressive resource for fleshing out the particulars of the attitudes that fortify the separation thesis, explains how these earlier descriptions matured. Much of the vision of the urban wilderness can be attributed to the runaway white middle-class imagination, absent (according to Davis) from any firsthand understanding of the inner city, so that any real threat is seen only though a "demonological lens."[21] This lens seems to be the same one that informed early colonial perceptions of the "natural" wilderness and its inhabitants. In both cases the perception of a classically savage wasteland was enough to designate an area as wilderness and justify its separation without a fully informed description of the physical place.

While the justification for such descriptions may stem from actual dangers in these areas, the results of this perception on urban policy are nonetheless extraordinary. Davis calls our attention to what he names "Fortress LA," a stronghold constructed against the new urban frontier in an attempt to create a new urban center in L.A., one that would be attractive to what *Urban Land* magazine calls "respectable people":[22]

> The carefully manicured lawns of LA's Westside sprout forests of ominous little signs warning: "Armed Response." . . . Downtown, a publicly-subsidized "urban

renaissance" has raised the nation's largest corporate citadel, segregated from the poor neighborhoods around it by monumental architectural glaciers. . . . In the Westlake and San Fernando areas the LA police department barricades streets and seals off poor neighborhoods as part of their "war on drugs." In Watts . . . [a recolonization] of the inner-city retail markets: a panopticon shopping mall surrounded by staked metal fences and a substation of the L.A.P.D. in a central surveillance tower.[23]

Indeed, the separation has been pursued to such an extreme that there is no pedestrian access to L.A.'s new revitalized downtown business and arts district. The implication of this is that residents of the largely Latino neighboring communities cannot walk to this area without the risk of crossing a highway.

Davis traces the fortress mentality at least partially to the plans for "re-segregated spatial security" outlined in the 1965 McCone Commission Report ("Violence in the City—An End or Beginning?"), which followed the original Watts riots. In the wake of the more recent riots, Davis traces these transformations more specifically as a deliberate movement responding to the "surrender" of the cities by a decade of Republican urban policies (with Democratic complicity) designed to privatize the public sphere. This is a return, he claims, of the cities to the "Darwinian or Hobbesian wilderness."[24]

I find this last language telling and more appropriate than the descriptions Davis gives of the fortress mentality in other terms. For instance, he suggests that an "urban cold war" is being waged in the city, specifically in the policy of "'containment' (official term) of the homeless in Skid Row."[25] However, it looks like a kind of dehumanization is going on here that was never the case for the teeming masses "trapped" behind the Iron Curtain. Cold War containment was a policy directed against the leaders of the Soviet Union and the Eastern bloc and not really against the people, who after all were supposed to be yearning to be free capitalists.

What is going on in this form of separation is a deeper anomaly. It is almost as if the classical wilderness is the object of preservation of legislation such as the Wilderness Act. Rather than a pseudoromantic view that the wilderness should be set aside because its good qualities deserve protection, the L.A. separation policy seems to implicitly argue that once we have conquered the classical wild we should preserve it as a quaint reminder of the environment to which we have not succumbed. (Or probably more consciously, we should encircle it because it is beyond hope.) There is no Sinclairian reform instinct here; "we" would not even think of allowing "them" into our areas, so we ought to just control this wild place and keep it separate as best as we can and as long as is possible.

One can see P2 and P3 in the not-too-distant background of this discussion of P1, but it may be interesting to look at them separately as well. Particularly in the language of the Rodney King trials, the ensuing riots, and the response, the assertion of the savagery of the natives of the urban wilderness and the superiority of those outside the inner city is striking.

P2 and P3 were already clearly present in the language of the first trial, which catalyzed the riots. In Robert Gooding-Williams's account of the rhetorical strategies of the defense attorneys (which occurs in a discussion very different from this one) the language of classical wilderness is thick:

> After inviting jurors to see events from the point of view of the police officers, the defense attorneys elicited testimony from King's assailants that depicted King as a bear, and as emitting bearlike groans. In the eyes of the police and then again in the eyes of the jurors, King's black body became that of a wild "hulk-like" and "wounded" animal, whose every gesture threatened the existence of civilized society. Not surprisingly, the defense attorneys portrayed the white bodies which assailed King as guardians against the wild, and as embodying a "thin blue line" that separates civil society from the dangerous chaos of the essence of the wild. (. . .) This animal, claimed one of the jurors, echoing the words of defense attorney Michael Stone, was in complete control and directed all the action. Still, somehow, the forces of civilization prevailed, preserving intact human society as we know it.[26]

Many racial, material, and psychological explanations could be advanced to explain the predominance of this language in the case. But it seems wrong to present any one of these as the sole explanation, just as it seems wrong to attribute the riots that followed exclusively to either wanton license or revolutionary fervor, as some have tried to do. Certainly, given the legacy of the naturalization of the urban wilderness as a cognitive component of outside views of the inner city, any explanation of such language should not omit the importance of its consistency with previous similar descriptions of the savagery of inhabitants of the wild.

In support of this claim, we should not find it surprising that the overwhelming response to the L.A. riots was to continue to view the inner city in the context of P1, P2, and P3. Even under the direction of a new, kinder, and gentler police captain, abuses (including possible unjustifiable murders) continued in the L.A. Police Department against inner-city residents. Youth gangs have been solidified in the public mind as the savage "enemy within" the city consistent with a distanced identification of urban wilderness. Separation efforts were redoubled in an attempt to protect threatened white sub-urbs at the insistence of the postriot Webster commission in what has been called the "Ulsterization" of the city. Successful arguments were made in the California state legislature to justify increased criminal penalties to further ease the control of urban savages, while aid to rebuild the city remains at a standstill.[27] A consensus seems to be forming in the business community that the region as a whole is slipping backward into a "neo-Disney, plastic Stone Age."[28]

But worries about these effects, even if they are attributed to a new urban wilderness, can be easily ignored. A simple law-and-order stance is all that is needed to argue that just as the idea of the classical wilderness was in part motivated by real dangers "out there," characterization of an urban wilder-ness is justified due to the actual dangers (soberly perceived) of life in Amer-ican cities.

There is an additional side to this story, however, that deserves brief men-tion and that may turn the heads of advocates of such a law-and-order posi-tion. There is some weighty evidence that natives of the urban wilderness have internalized this designation as "savage," contributing at least to de-structively low self-esteem and at worst, to increased violent and threatening behavior. This worries me, not because I can show that there is a causal rela-tionship here, but simply because it is a very uncomfortable correlation.

One disturbing case occurred in New York on April 20, 1989, when a young, middle-class white woman was severely beaten and raped in Central Park by a gang of young African Americans who themselves described what they were doing as "wilding." Newspaper headlines around the city immedi-ately capitalized on this language of the wilderness and the sense that the lines of separation had been crossed: "Teen Wolfpack Beats and Rapes Wall Street Exec on Jogging Park," "Central Park Horror," "Wolf Pack's Prey," "Fe-male Jogger Near Death After Savage Attack," "Park Marauders Call It Wilding."[29]

Tremendous controversy ensued in the city over the case, with accusa-tions of racism and reverse racism bantered about in the media. Calls for more spatial separation (P1) were stimulated by a demonization of the young men accused of the crime (P2) while the media held up the victim as a model of the resistance of middle-class morals and steadfast refusal to sur-render territory to the beasts of the inner city (P3)—even to the point of maintaining the right to jog in Central Park at night. But what strikes me most of all is the way that the language of the classical wilderness had seeped into the urban setting, helped to consciously self-designate the behavior of inner city residents, and *possibly* contributed to their ability to perform such an act. At least it is clear that the young men had been affected by constant cultural cues that they were inferior, somehow less than (civilized) human.

Yusef Salaam, one of those convicted in the case, reveals something of this problematic in a rap he presented as part of his personal statement in court before he was sentenced:

> I'm kind of laid back, but now I'm speaking so that you know
> I got used and abused and even put on the news
> I'm not dissing them all, but the some I called
> They tried to dis me like I was an inch small, like a midget,
> a mouse, something less than a man.[30]

Here may lie part of the concrete legacy of the classic view of the wilderness and its inhabitants that we can do without.

IV

For some time now we have been in the midst of a transition, a drift in meaning from what I have been calling the classical metaphor of wilderness to a romantic view. Part of the reason we are moving toward an identification of wilderness as a thing to be preserved, saved, cherished, and not feared is that the conditions that caused us to vilify the wilderness are by and large no longer sustainable. For scientific, philosophical, anthropological, and other reasons it does not make much sense to adhere to P1, P2, and P3 in our cognitive construction of wild nature.

Nonetheless, as I have argued here, a remnant of classical wilderness persists as the urban wilderness, so the three theses are still doing some work, and possibly harmful work at that. In the urban wilderness the cognitive dimension of the wild as that place where the savage lesser-human exists and where civilization breaks down remains. Here civilization identifies and demonizes the new savage while under the influence of romantic wilderness, it may at the same time idealize and romanticize the old savages (namely, indigenous peoples).

Perhaps, though, the inner city is the last refuge of classical wilderness. Perhaps (I hope) it is the last gasp of demonization of those we find different from ourselves. Its death may also mark the end of the vilifying of an internal boundary between what we wish to be (virtuously rational) and what we fear to become (savagely passionate), at the expense of others.

Urban wilderness should eventually pass into history, just as the original classical wilderness is passing. But we ought not to stop with that, for the transformation from classical to romantic views of wild nature took much time, and the persistence of the classical view contributed to a lot of suffering. Even if the transformation out of urban wilderness is inevitable, that does not mean it could not be hurried along. Like the Renaissance astronomers, one of our first important tasks is to accept that our object of study is up for grabs, that anomalies exist that need to be explained, and that there may be some powers (in the astronomers' case it was the Church) that have an interest in falsely naturalizing the object of study. Once these important principles are recognized, transitions can be pushed along—if not toward ultimate understanding, then at least toward recognition of mistaken conceptions. Once we begin investigating the idea of wilderness we can see that we are not necessarily wedded to any one view.

Finally, consistent with this argument, I do not see a need to further philosophically define wilderness, since the word may never have picked out something approximating wild nature at all. The expression "wild nature" (or suitable variations to designate specifiable gradations in kinds of wild nature) ought to do the work without the genealogical baggage of "wilderness." But for political reasons the term is still meaningful and needs to be negotiated around, particularly for those theorists who want to communicate with policymakers.

With classical wilderness as a referent to wild nature pretty much out of the picture, one alternative is to work with the romantic idea of wilderness. But this is unsatisfactory because romantic wilderness is no less a metaphor for something being projected onto wild nature by human desires than is classical wilderness (though possibly with less potential harm to others). Also, it appears that romantic wilderness is slowly eroding with time and so requires some work for comprehension.

Instead, for the purpose of communicating with those who still require the term, something on the order of an indexical definition of wilderness is needed. This, simply put, is a sense of wilderness out of which we can designate an area as wilderness by pointing at it and saying, "this is wilderness" or "that is wilderness" and it ought to be preserved as best as we can. I call this view the "thin" conception of wilderness.[31] Thin wilderness is just wild nature, but that does not mean that it designates a shallow view of the wild.

When we use the term *wilderness,* even thinly as I suggest, it still marks out something special in the world. The term may still carry theoretical weight and it may still serve as the basis for sophisticated normative principles. But such a term need not carry the baggage accumulated over time by the classical or romantic view of wilderness and may be constructed coherently without necessarily *needing* to refer to some ideological cognitive status of the person using the term. Such a term would, I hope, be more flexible, more malleable, and hence less subject to abuse. Without a catholic designation of what the term must refer to and what qualities it must convey to the thing it describes, there is greater possibility of keeping it from being naturalized in inappropriate contexts. There is in short, less possibility of using the designation of wilderness as a conduit for harm.

Notes

1. See John Rennie Short, *Imagined Country: Environment, Culture and Society* (New York: Routledge, 1991), 6.

2. Ibid.

3. Short points out that the two views do of course share certain commonalities. In English, for example, *fear* and *revere* both mean "to stand in awe."

4. See Short, *Imagined Country*, and Roderick Nash, "The Romantic Wilderness," in his *Wilderness and the American Mind*, 3rd ed. (New Haven, Conn.: Yale University Press, 1982), 44-66.

5. For a history of this argument see Roderick Nash, *Wilderness and the American Mind*; Joseph Sax, *Mountains without Handrails: Reflections on the National Parks* (Ann Arbor: University of Michigan Press, 1980); and, for an account by one of the architects of this vision, Frederick Law Olmstead, "The Yosemite Valley and the Mariposa Big Trees, a Preliminary Report (1865)," *Landscape and Architecture* 43 (1952): 12-25.

6. There are many very good studies of such positions. For one account of a few see Andrew Light, "The Role of Technology in Environmental Questions: Martin Buber and Deep Ecology as Answers to Technological Consciousness," *Research in Philosophy and Technology* 12 (1992): 83-104.

7. Short, *Imagined Country*, 9.

8. See Ronald Takaki, *Iron Cages* (Oxford: Oxford University Press, 1988), especially chapter 2.

9. Short, *Imagined Country*, 9.

10. See, for example, the excellent genealogy of the idea of wilderness in Max Oelschlaeger, *The Idea of Wilderness: From Prehistory to the Age of Ecology* (New Haven, Conn.: Yale University Press, 1991), especially the account of the idea of wilderness in the Middle Ages.

11. One example that I will not treat here is the claim that peasant women in El Salvador do not have a conception of wilderness at all, let alone one that fits either the classical or the romantic conception. See Lois Lorentzen's "Bread and Soil of Our Dreams: Women, the Environment and Sustainable Development—Case Studies from Central America," in *Grass Roots Ecological Resistance: The Global Emergence of Radical Environmental Movements*, ed. Bron Taylor (New York: SUNY Press, forthcoming 1995).

12. See Nash, *Wilderness and the American Mind*, especially the remarks by the Ogalala Sioux Standing Bear on this point, xiii.

13. Ibid., xiv.

14. Thomas Blackburn and Kat Anderson, "Managing the Domesticated Environment," in *Before the Wilderness: Environmental Management by Native Californians*, ed. Thomas Blackburn and Kat Anderson (Menlo Park, Calif.: Ballena Press, 1993), 18.

15. Ibid., 19.

16. See, for example, Robert A. Wood, *The City Wilderness* (Boston, 1898).

17. See for example *"Yours for the Revolution": The Appeal to Reason 1895-1922*, ed. John Graham (Lincoln: University of Nebraska Press, 1990). The image of the dark urban factory environment can be seen in political cartoons as well as essays. For example, one cartoon in *The Appeal* from 1922, "The Trap," depicts a brutish, oversized, apelike capitalist luring workers into a factory entrance that is actually the entrance to a huge forbidding forest. Importantly, Sinclair's *The Jungle* appeared as a serial in this publication before it appeared as a book.

18. Upton Sinclair, *The Jungle* (New York: Bantam Books, 1981), 40-41.

19. Short, *Imagined Country*, 25, 26. Nash briefly mentions this transition to an urban wilderness in *Wilderness and the American Mind* (see xii, 3, 143-44) but does not go so far as to explain the sense in which this transition depends on the classical conception of wilderness in an exclusively cognitive context. I hope that my discussion will deepen the few references in Nash's book as well as update their significance.

20. Mike Davis, *City of Quartz* (New York: Vintage Books, 1992), 224, provides a summary of the National Committee on the Causes and Prevention of Violence, final report, *To Establish Justice, to Ensure Domestic Tranquillity*.

21. Ibid. Another good source for the politics and ideology behind urban geography is Edward Soja, *Postmodern Geographies* (London: Verso, 1989).

22. N. David Miller, "Crime and Downtown Revitalization," in *Urban Land* (September 1987): 18, cited in Davis, *City of Quartz*, 231.

23. Davis, *City of Quartz*, 223.

24. Mike Davis, "Who Killed L.A.? A Political Autopsy," *New Left Review* 197 (1993): 3-28.

25. Davis, *City of Quartz*, 232.

26. Robert Gooding-Williams, "Look, A Negro!" in *Reading Rodney King/Reading Urban Uprising*, ed. Robert Gooding-Williams (New York: Routledge, 1993), 166. Gooding-Williams in part drew the record of the trial from the *Los Angeles Times*, 3 April 1992, 21 April 1992, and 22 April 1992.

27. See Mike Davis, "Who Killed Los Angeles? Part Two: The Verdict Is Given," *New Left Review* 199 (1993): 29-54.

28. Ibid., 47.

29. Headlines reprinted in Joan Didion, "Sentimental Journeys," in *The Best American Essays 1992*, ed. Susan Sontag and Robert Atwan (New York: Tickner and Fields, 1992), 6.

30. Ibid., 9.

31. I am drawing this designation from Steve Cullenberg's argument for a "thin" definition of socialism—another term with a controversial past. By designating such a conception of socialism, Cullenberg hopes to create the intellectual space to use *socialism* as a referring expression, retaining its theoretical use without drawing in its past ideological baggage and utopianistic expectations. See Cullenberg, "Socialism's Burden: Toward a 'Thin' Definition of Socialism," *Rethinking Marxism* 5 (1992): 64-83.

HEAD WATERS OF SACRAMENTO RIVER

64 Crossing S. P. Co. Oregon Division

Epilogue: Paradox Wild

David Rothenberg

I got up that morning in the wet Brooklyn rain with this one idea: Tonight I will touch the Pacific Ocean. And then the description of the journey follows—the planes, the waiting, the cars, the sounds, the call of the wild, the waning light, the roar not placid in the least, the archaic image of the West (that is the wild or is not?)—and at last the resolution.

"The West is another name for the Wild," announced Thoreau. Enough writers since have quoted this phrase, cast it anew and taken their own journeys from civilization toward the frontier. And that frontier recedes. Each year it seems more of a lie, and the expanding tendrils of an information society work as well on mountain paths as on superhighways.

Then again, the very word *West* has gone on to other connotations, meaning this culture that has won the battle, expanding everywhere, freedom winning over restraint. Capitalism becomes everyone's religion, and we fortunate decide to help the rest of the world and tell them how to live. Is *this* the place that is another name for the wild? This craziness, this society spinning out of control, now as dangerous as the untrammeled territories we have nearly subdued.

Now we see the word *wild* is an enigma, a riddle neither good nor bad. It's sexy and dangerous; we want to save it, yet we can't fence it in. Whatever we decry, the word will not lose its range of allusions. It won't just be what the primitivists want, or the wilders of the urban ghettos. It's a concept that comes only from civilization, and means nothing outside of a human take on the world.

I've moved to New York for awhile, and I've come to see nature differently as a result. Every tree is to be cherished. Every voyage away, even but an hour, seems a trip to a quiet oasis where the world shouts its verdancy to me, the value of trees, the green of the forest, and the blue of the sea. If I go far enough away, especially north, I'll reach something that sings its heart out as wilderness. But the terror, the danger, the *adventure* of the wild, that's down the avenues of home. Two miles away. Spike Lee calls it Crooklyn.

What place have these ruminations in a book on wilderness, that sacred goal far off in the Louisiana Purchase? We must seek and then follow our

wildest ideas. I went West to find an end to this volume, some image to conclude it.

Wandering through the Helmut Jahn United Terminal at O'Hare, en route
to the wild. The causeway from the main terminal to the satellite is a fragment of a theme park, an Epcot future prototype adventure, a complete arti-
ficiality but composed completely out of the danger and brightness of nature. Colors from throughout the spectrum are gridded on the walls in soft
shades like those gracing the covers of software manuals, above in flashing
neon jaggies, lightning bolts showing America on the move, heading west to
east, east to west to frontiers of change, technologies, and personal escape.
This is the wild right here, my friends, two flat escalators in either direction,
walk left, stand right, fragments of electronic celestes and pianos adding
open, mixolydian inflections, a music with no beginning or end, harbinger
of the hope of travel, the possibility of discovering some nature still and always beyond our reach. A pause in the midpoint, the lyrics, the mantra of
the experience. "The walkway is about to end. Please slow down." We don't
even need these words to guess the two-faced message of the happening
around us: bright lights, on and off—GO! Soft sounds, restful subliminals,
slow down, take a break, rest, enjoy—STOP! Just for a while. It's taking us
from plane to plane, from periphery to center, from home to away, from the
city to the wilderness. It's only the latest conduit, a people mover of the moment, offering travel with as much paradox and adventure as ever, though so
mediated by machinery. The pandemonium is wild, it's confusing, I'm on
my way away. But the solace, the dream, the peace . . . I am carried through
this installation, this performance piece, this illusion, further on the path to
nowhere.

See? It's so much easier to write or to talk about machines than it is to describe nature. Bill McKibben spent twenty-four hours watching TV, then
twenty-four hours backpacking in the woods behind his Adirondack home
in order to elucidate the contrasts possible in our modern world.[1] As fine a
writer as he is, McKibben has more to say about loud TV culture than about
silent nature. Once we have found the wild, it remains hard to speak of it.

At nightfall I was crossing miles of orange dunes to discover the sea. This
was a rare maritime desert, a strange and beautiful landscape, at the other
edge of the continent from where I began the day. The sand extended as far
as I could perceive to the north and the south, while west lay the setting sun,

the horizon of water, and beyond, paradoxically the civilization of the *East*, coming closer to our borders, an alien world neither Old nor New.

The dunes were bathed in a golden light, and the ephemeral lakes were like jewels among mountains of sand. My footprints were not much of an intrusion, because the wind would blow them over in a day or so. Nothing is fixed in a landscape still being created by air.

Nevertheless, I embraced the quiet thousands of miles from home, knowing it was a sixteen-square-mile bounded area. Safe from the dune buggies that raced up and down the rest of these beachy expanses. I heard them in the distance, and laughed. What kind of wild place was this?

On and on toward the water, soon stuck up to my knees in a swamp. The ocean, the roaring, crashing Pacific Ocean, was only a short distance away. I did not make it that time, hearing noises natural and human. Soaked to the gills, I turned around and headed East. My uncertainty is another name for the Wild.

Skiing atop twelve feet of old, mossy snow beneath Mt. Hood in Oregon, I see the following words on an aluminum sign:

WELCOME TO WILDERNESS
The Forest Service is maintaining this area for you
by providing only a few primitive trails, bridges, and signs.
Your visit may include a degree of challenge and risk.

These lines are drawn everywhere, bounding the wild from the tame. The government abdicates responsibility, and allows me to proceed into wilderness at my own risk. It takes no blame for me now, yet silently still offers rules. What can or cannot be done. It is a tame place, this demarcated wild. Outside the line can be any manner of ugliness, from an eyesore of a factory to a mountainside stripped of its trees. Then there's a marker: let there be wilderness. Nature is wholly separated from the human project. There's the biggest danger in the wild idea—it assumes that nature is best served by booting humanity out.

Our writers here have sought to use language carefully, yet without losing the ambiguity of "wild." Like the best of words, it eludes easy explanation. Just saying it, using it, trying to understand its function in a speech or a song makes a profound comment on the human place in nature. And its meaning is changing, much as human proliferation is continuing to encroach upon a world that we want to see only as resource.

In an antique store on the Olympic Peninsula, I find, mysteriously, an anthology on these same problems, *Wilderness: America's Living Heritage,* edited by the indefatigable David Brower in 1961.[2] Many of the same issues are discussed, from wilderness in the arts to wilderness in urban sprawl. It is an intriguing book, from what seems a kinder, gentler time. All the authors are white, generally wealthy, American males. The unity of the cultural dream is never questioned. Heroes such as Justice William O. Douglas and Stewart Udall cry out that the time is running out, that we must save the little we have. Note that this is before the Glen Canyon Dam, before the sixties, before the crises of global development that caused this Wild Western civilization to begin to question its roots.

Yet the rhetoric is so similar, the concern is so heartfelt, so solidly American, for better or worse. I see David Brower on the streets of Eugene; he's talking the same talk, walking the same walk. Why not? We're still not, by and large, listening. I hope I'm not naive to think that as America recognizes its internal diversities, respect for nature and some deference to the power of the wild will continue. I am inspired and only a bit surprised that my students at the New Jersey Institute of Technology in Newark nearly all seem to accept and defend the idea that nature has intrinsic value, independent of whether it is useful for human purposes. However problematic that concept may be to argumentative philosophers, it makes intuitive sense to more and more people. That is something that has changed since 1961. Heavily publicized disasters have spurred on this realization, as have the general yellowing of the air, the congestion on the freeways and in the head.

And the wild, is this respected, or still feared? Wilderness today must be seen from many contexts: part of the diversity within America, and of America within the world. It cannot be simply exported across the globe as another imperialist policy. It cannot be reduced to a boundary, or to a term identifying a specific human use of land. *Wild* is a power word, a beguiling force, both a romantic and a classical concept. It will elude explanation, and we will continue to fight to explain it, and to evoke it. We will practice the wild in the midst of the tame, keeping civilization always on its toes. The wild will matter to those deep in the inner cities, and to those who never set foot in a car. It will still be both good and bad, or neither, a challenge to the best laid plans.

Andrew Light and I take the final leg of this journey together. Driving northwest through the rain. Looking for the wild. *Light.* Good name for a philosopher, conjuring images of bright ideas. Accurate. He's like that, too. Exact, discriminating. I tend to wander off, to look for answers in reverie. "Nature," he cautions me, "should not become religion." Right he is. No place out there or under our muddy feet has the answers. It is greater than us, and great enough to be indifferent. We will have to figure out our own rules, and imagine that the world, too, wants to play.

Walking through the Olympic mist, following a marked path to a big tree. As if all the trees around aren't big enough. They are growing and dying, collapsing and returning to the underbrush of the temperate rain forest, soaking wet. We come to a place of wild design, a precise but human-made landscape: a long tree felled to a pathway, gridded so that we will not slip. It crosses a slow-flowing, straight river, almost a full liquid mirror of the straight line of the tree. Wood crosses water, we walk across. It is a human landscape that bespeaks nature, like a Zen walk, tree tracking stream.

I am drawn to these places, where a human action, noticed, created, built, or just drawn, allows the traveler to step one step closer to the interior of nature. I believe in this kind of cultural leap even more than the wild, for it suggests that there might be a way to accept human dwelling into the natural world without setting up these barbed-wire fences between wilderness and the less-regulated national forest, "Land of Many Abuses."

I am easy. A place of simple experience convinces me it can be done. One path in the forest. One leap takes you there.

We hear that far into the wild country, near the end of the path to Mt. Olympus, someone sits on the glacier with a laptop, keystroking away. "I'm here! I'm connected! Here on this mountain I'm just one exit off the information superhighway . . . " Is the machine in the forest a threat, or the answer? Will we live lightly on the earth in the future, each able to find a personal place in the wildernesses of our choice, or will a virtual life leave us so out of touch with nature that we will no longer know what is real, what is mediated, and what is wild and free?

On the coast, in the storm, we are running on the sand, trying to beat the tides. The rush of wind. The *sound.* The close inhospitability of it all. The immediacy of the wild that is both frightening and exhilarating is the old aesthetic category of the *sublime.* There cannot be total safety. There may be some risk in these activities. That's not strong enough for advice: there *will* be risk. You must put your life on the line to discover the wild. You must not be afraid to doubt, to throw away all presuppositions if they hold experience back. It may not be good for you, but it will encourage the *wild*. And this wild isn't everything; no, it harbors, along with creation, total annihilation if we lose all control.

By the Portland airport the shuttle-bus driver tells me Brooklyn is his hometown. He loves the mountains, the calm in the people who gaze daily on volcanoes and can breathe the scent of the forest even on Main Street. "Sure, sure I like to go back. But hey, be careful. It's a *jungle* back there." And I step onto the concrete sidewalk. Then get on the plane, sit still for hours, watch the clouded wild lands of America beneath me, tense before the descent.

Notes

1. Bill McKibben, *The Age of Missing Information* (New York: Viking, 1992).

2. David Brower, ed., *Wilderness: America's Living Heritage* (San Francisco: Sierra Club Books, 1961). This volume contains essays, speeches, and conversations by Ansel Adams, Pat Brown, William O. Douglas, Joseph Wood Krutch, Sigurd Olson, Gerard Piel, Wallace Stegner, Howard Zahniser, and other pioneers in the raising of American wilderness consciousness.

Contributors

David Abram, philosopher and magician, is the author of the forthcoming book *The Earthly Body of the Mind: Animism, Language, and the Ecology of Sensory Experience.* He is a recent recipient of a Rockefeller Fellowship in the Humanities.

Douglas J. Buege completed a doctoral degree in philosophy at the University of Minnesota with a dissertation in environmental ethics. His current research examines relations between native peoples and the environmental movements of Europe, Canada, and the United States.

Denis Cosgrove, professor of human geography at London University's Royal Holloway College, has written extensively on meaning and symbolism in landscape and the cultural construction of wilderness ideals. His books include *Social Formation and Symbolic Landscape, The Iconography of Landscape,* and *The Palladian Landscape.* He is editor of *Ecumene,* a journal devoted to themes of environment, culture, and meaning.

Robert Greenway is a professor of psychology at Sonoma State University, currently on leave and living on the Olympic Peninsula, collecting his thoughts on directing a university wilderness program for over twenty years. His research concentrates on developing an ecopsychology: adapting the paradigms of psychology to explain what happens to us when we reach out and make contact with the surrounding natural world.

R. Edward Grumbine directs the Sierra Institute Wilderness Field Studies program at the University of California Extension, Santa Cruz. Since the mid-1970s he has been working at the craft of teaching university courses in the backcountry of the western United States.

Marvin Henberg is dean of the faculty of Linfield College in McMinnville, Oregon. He is currently writing a book on the ethics of wilderness policy. When not at work he's up in his cabin by the Three Sisters in Bend.

Irene Klaver, a native of the Netherlands, is finishing her doctorate in philosophy at SUNY Stony Brook and teaches at Montana State University, Billings.

Andrew Light is a postdoctoral research fellow in the Environmental Risk Management Program and the Department of Philosophy at the University of Alberta. He has published several articles on environmental moral and political theory and philosophy of technology; currently he is coediting an anthology entitled *Environmental Pragmatism.* He is cofounder of the Society for Philosophy and Geography.

Lois Ann Lorentzen teaches environmental ethics in the Department of Religion at the University of San Francisco. She is also a mountain climber, often to be found roaming the Sierra Nevada of California and the Oregon Cascades.

Max Oelschlaeger is an inveterate backpacker and wilderness trekker, author of *The Idea of Wilderness* and *Caring for Creation,* and editor of *The Wilderness Condition.* He teaches in the environmental philosophy program at the University of North Texas during the academic year, and inhabits Colorado's San Juan mountains during the summers.

David Rothenberg is director of the Program in Science, Technology, and Society at the New Jersey Institute of Technology. He is the author of *Hand's End: Technology and the Limits of Nature, Is It Painful to Think? Conversations with Arne Naess,* and numerous professional and popular articles on the human place in nature. He is also a composer and jazz clarinetist, with several recordings available. He lives in New York, a place that seems at times wilder than anywhere else.

R. Murray Schafer is a composer and researcher into human awareness of the soundscape. His book *The Tuning of the World* is the pioneer work in sonic environmentalism. His music and theater pieces have received numerous awards and have been performed all over the world. He lives in the forest somewhere outside Toronto.

Tom Wolf learned most of what he knows about wild ideas from Holmes Rolston at Colorado State. The rest he learned near his home at the base of the Sangre de Cristo Mountains, and over the Sangres in the San Luis Valley at Colorado College's Baca Campus, where Wolf teaches ecology. His essay here comes from a forthcoming book, *Colorado's Sangre de Cristo Mountains.*

Index